# GROWING FRUIT IN YOUR BACKYARD

*Lee Reich*

## BURPEE

A Burpee Book

Macmillan • USA

**MACMILLAN**
A Simon & Schuster Macmillan Company
1633 Broadway
New York, NY 10019

MACMILLAN is a registered trademark of Macmillan, Inc.

Library of Congress Cataloging-in-Publication Data
Reich, Lee.
    Growing fruit in your backyard / Lee Reich.
        p.    cm.—(Burpee American gardening series)
    "A Burpee book."
    Includes index.
    ISBN 0-02-860987-5
    1. Fruit-culture.    2. Fruit.    I. Title.    II. Series.
SB355. R455    1996
634—dc20                                                    95-35878
                                                                  CIP

Printed in the United States of America

10 9 8 7 6 5 4 3 2 1

BOOK DESIGN BY NICK ANDERSON

# *Introduction*

What better source for enticement in the Garden of Eden could there have been than a crisp apple? Or any fruit, for that matter, plucked fully ripe, with the coolness of night still lingering in its flesh and its skin just warmed by the kiss of morning sun. Fruits are the most delectable offerings from the garden.

*Bowl of raspberries, black currants, red currants, and gooseberries*

As a backyard fruit grower, you have an almost unlimited selection of fruits to choose from: thousands of varieties of apples or pears, for example. If certain varieties are not available locally, mail-order sources can ship them to you. Or you can try your hand at a propagation technique such as grafting and "make" your own plants. You can choose varieties whose flavor you consider or hope to be best-tasting, perhaps at some sacrifice to productivity or beauty. These are considerations in the selection of commercial fruit varieties but are not necessarily all-important for backyard fruits. 'Ashmead's Kernel,' for example, is an apple that is too ugly to sell, but it is tops in flavor. You also might choose to grow a fruit such as gooseberry or pawpaw, neither of which you are apt to find on grocers' shelves.

Many gardeners grow flowers, shrubs, and vegetables but consider fruits too difficult to grow. Fruit growing presents a challenge to commercial growers, who must eke maximum yields from every branch of a hundred acres of trees. Growing fruit in the backyard, however, can be highly enjoyable and not difficult at all if you begin by selecting appropriate varieties and types of fruits for your climate and site.

*Basket with many varieties of apples*

Not to be overlooked is the beauty that fruit and berry plants add to the landscape. A peach tree in blossom is a cloud of pink, its beauty rivaling that of ornamental cherries. The red stems of blueberry bushes enliven a winter snow scene, as do those of red-twig dogwoods. And elderberry stems, bowed downward by their heavy clusters of white blossoms, are reminiscent of viburnums (to which they are related). But whereas the ornamental cherry, red-twig dogwood, and viburnum are grown solely for their flowers, peach, blueberry, and elderberry are ornamental *and* yield luscious bounty.

So go ahead and consider planting fruit trees for beauty and for food. The pages that follow will guide you in the selection of fruits to grow and where to plant them, as well as in such details as pruning, feeding, and watering your plants. With a reasonable amount of time, labor, and study, you can be harvesting apples—as well as plums, grapes, kiwifruits, and more—from your own little Eden.

# Chapter 1

# THE FRUIT GARDEN PLANNER

## SITE

*Match the fruit plants you want to grow to the conditions in your yard, and your plants will be healthier and bear more fruit. In some cases the site might even affect the flavor of the fruit.*

Surprisingly, perfect soil is not all that important in site selection. It would be important if you were planting acres and acres of orchard. But for backyard fruit, you can feasibly, if necessary, modify the soil right where you plant to suit the needs of a specific tree, bush, or vine.

Two of the most important considerations in determining where to plant and what to grow are sunlight and winter cold. Most, but not all, fruit plants need full sunlight—six or more hours of direct sun per day. If your yard lacks even this much sunlight, you can still go ahead and plant, but expect some sacrifice in yield and keep an especially close eye out for diseases. As an alternative, plant fruits such as gooseberries and currants, which can grow even in partial shade.

Fruit plants vary in how much cold they can tolerate. Low temperatures might kill just the fruit buds (in which case you lose the forthcoming season's crop), branches, or even the whole plant. For a general idea of the average minimum temperatures of winters where you live, consult the USDA Plant Hardiness Zone Map at the back of this book. As you look over nursery catalogs, deciding which fruits to plant, note the hardiness zones listed for the fruits. Peaches, for example, are generally adapted from Zones 5 to 9, corresponding to regions experiencing average winter lows of −15° to +25° F. Specific varieties may have narrower adaptation; 'Desert Gold' peach, for example, is adapted only to warmer regions, Zones 8 and 9.

The USDA map, covering the whole country, cannot help but generalize over relatively large areas. If you are new to an area, speak to neighbors to find out how low the mercury usually plummets most winters in your neighborhood. Back up your own experience and that of neighbors with some hard facts—purchase a minimum-maximum thermometer. This type of thermometer, besides telling the current temperature, registers the minimum and maximum temperatures that were reached since the thermometer was last reset. If the mercury plummets some blustery winter morning to −20° F at 3 A.M., then "warms" to −5° F when you awaken at 6 A.M., you will still know just how cold it really was. (Your plants already knew!)

*Minimum-maximum thermometer*

Keep in mind that the actual low temperature in any winter will not necessarily be the same as the "average minimum" temperature. If you live in Vermont or Montana, do not let a string of atypically warm winters lull you into planting nectarines. Use the USDA map along with your own observations and those of neighbors.

Minimum winter temperatures are as important to gardeners in warm regions as they are to northern gardeners. This is because fruit plants that lose their leaves and go dormant in winter will not resume growth in spring until they experience a certain number of cumulative hours of cool—not cold—weather. Such "chilling" occurs for most plants between 30° and 40° F. Temperatures much higher or lower than this amount do not put hours into the cumulative chilling "bank."

The amount of chilling required varies with both the type and variety of fruit. Most apples, for example, need 1,000 to 1,500 hours of chilling temperatures before they will begin to grow again in spring. This is of little concern to most gardeners except those in areas with short, mild winters, such as in northern Florida or southern California. If your garden is in a region with short, mild winters, plant the types of fruits that need little or no chilling—such as figs, pomegranates, and grapes. If you must have apples or other fruits of colder regions, plant "low chill" varieties—'Beverly Hills' apple and 'Southern Flame' peach, for example, both of which require only a few hundred hours of chilling.

This chilling business also works the other way around, farther north. Where winters are moderately cold, the chilling requirements of low-chill plants are fulfilled early in the year. The result: overeager blossoms open too early and are usually nipped by subsequent frosts. Therefore do not plant low-chill fruits in the North.

Not all low-chill fruits come from warm winter regions; plants native to areas where winters are long and steadily cold also are genetically programmed to begin growth after only a short amount of chilling. In their native habitats, the chilling requirements for such plants are not fulfilled until late spring, at which time growth must begin quickly in order for the fruits to ripen within the relatively short growing season. But where winter temperatures fluctuate, such as over much of this country, the chilling hours for these plants slowly accumulate through the winter. As a result, plants such as apricot, which come from regions with long and steadily cold winters, bloom early, and their blossoms are often nipped by late frosts.

How cold your winters get may be the most important consideration about your climate for growing a particular fruit, but it is not the only one. If you live in a high-elevation or far-north location, where the growing seasons are short, or if you live along the ocean, where summer temperatures remain cool, you may also have to take your summer weather into account.

Rainfall is usually less limiting to what you can grow than is temperature. Most plants need a steady supply of water throughout the growing season, but you can water your plants if natural rainfall is inadequate (see pages 19–22). Pomegranates and sweet cherries tend to split if it rains as these fruits are ripening, but you can get around these problems with timely harvesting and by planting nonsplitting varieties.

## Spring Frost Hazard

Many fruit plants bloom in late winter or early spring, at which time their blossoms are vulnerable to subsequent subfreezing temperatures. Such frosts kill the blossoms, rarely the plant, but you lose the crop for that year. Because water moderates temperature changes, the areas least likely to experience damaging late frosts are those near oceans, large lakes, and rivers. Elsewhere, help your plants avoid the effects of late frosts by utilizing microclimates.

A microclimate is a small area where the climate is different from the general climate. If you've ever warmed yourself against a sunny south wall on a winter day, you've experienced the effect of a microclimate. The wall caught and held extra heat, keeping you comfortable even though the air a few yards away may have been below freezing.

To help a fruit plant avoid damaging frosts, find the opposite type of microclimate as that sunny wall— one that keeps a plant cool, thus delaying blossoming until all frosts are probably past. North-facing slopes are one such microclimate because they trap less of the sun's heat than does level ground or south-facing slopes. Similarly, you can plant an early-blooming fruit tree near the north side of your home or garage. There, your plant will be shaded from low winter sun. Just make sure you plant far enough away from the wall so that summer sun falls on the plant.

Late winter and early spring freezes commonly occur on still nights when the sky is clear. Under such conditions cold air, being heavier than warm air, hugs the ground and flows downhill, filling up low spots, as would water. Avoid spring frost damage by not planting in low areas, especially if you are growing a low-growing, early-blooming plant such as strawberry.

Because cold winter winds sweep across the tops of hills, the ideal location for fruit plants is right on a slope rather than at the top or the bottom. Make sure that no obstructions, such as dense hedges or walls, stop the flow of cold air down a slope.

Do not forsake the pleasure of growing fruit for lack of a site free from spring frosts. Fruit plants such as persimmons and grapes bloom after all chance of frost is past. Some fruits—gooseberries, generally, and 'Wagener' apple, specifically—have blossoms that are somewhat resistant to frost. And you can easily drape a blanket over a strawberry bed on a night when frost is predicted and the plants are in bloom. Or grow plants in pots (see pages 31-34). How about a greenhouse?

## Stretching the Limits

The ideal in fruit gardening is to perfectly match plant and site conditions, but gardeners seem always to be trying to stretch the limits of what they can grow. The gardener in New York tries to grow figs, the gardener in Georgia tries to grow citrus, and the gardener in southern California tries to grow apples. In each case, the plants are unsuited to the climate, but something about the challenge of growing these fruits and savoring their tastes spurs such gardeners on.

Microclimates, once again, make such endeavors feasible. Where a plant is particularly susceptible to diseases, choose a site with unobstructed breezes and plenty of sunshine. Where extra heat is needed in winter, set a plant on a south-facing slope, against a south wall, or near paving. (Watch for and protect leaves and blossoms that appear too early, though.) If more cold protection is needed, wrap a plant in burlap for the winter. Or build a wire cage around the plant and stuff it with autumn leaves. For even greater cold protection of plants with flexible stems, dig a trench right next to the plant and bury the stems for the winter. Or grow a plant in a pot, which you can move to warmth when necessary.

On the other hand, if your plants need cooler summer temperatures than your region generally offers, utilize cooler microclimates such as partly shaded sites, as well as those on north-facing slopes or near the north sides of walls or buildings.

# POLLINATION

Aside from providing good site and growing conditions for your fruit plants, you may also have to provide for pollination. Pollination is the transfer of pollen from the male parts of a flower (the anthers) to the female parts of a flower (the stigmas). Within the flower, the pollen cells unite with the egg cells, forming a seed and, in so doing, stimulating development of the fleshy covering around the seed—the fruit!

Pollen can stimulate fruit formation only if it comes from a flower of the same type of fruit. Thus, an apple flower can be pollinated by an apple flower but not a strawberry or pear flower.

A flower may have male or female parts or both. If male and female flowers are separate but on the same plant, the plant is *monoecious* (from the Greek words meaning "one house"). A *dioecious* plant (from the Greek words meaning "two houses") has either male or female flowers. Most cultivated fruits have what botanists call *perfect* flowers; that is, each flower has both male and female parts.

Not all plants with perfect flowers can set fruit with their own pollen. Flowers that need *cross-pollination* will not set fruit unless they receive pollen from a different variety of the same type of plant. Apples, for example, need cross-pollination, so a 'McIntosh' tree needs another variety, such as 'Red Delicious,' nearby in order to bear fruit. The 'Red Delicious' will similarly bear a crop using pollen from the 'McIntosh' flowers.

Certain varieties of fruits may have special pollination needs. 'Magness' pear produces poor or little pollen, so it is incapable of pollinating any other pear. And although 'Bartlett' and 'Seckel' each produce good pollen, an innate incompatibility prevents them from cross-pollinating each other. In these cases, three different varieties are needed for fruit production. Some of the hybrid plums need specific pollinators, depending on their parentage. You will find this sort of information listed with the description of a variety in nursery catalogs.

The most obvious way to provide a pollinator, when needed, is simply to set another plant in the ground. Plants should be within a hundred feet of each other for effective pollination. Before you choose pollinators, also make sure that bloom times overlap. 'Gravenstein' apple, for example, blossoms very early and so would be a poor choice to plant with a late-blooming variety such as 'Rome Beauty.'

There are ways to get around having to put another plant in the ground when growing a fruit that needs another pollinator plant. Perhaps your neighbor is growing a plant that is a suitable pollinator. Perhaps pollen could be supplied by wild plants nearby. If you know of a suitable pollinator plant that is not nearby, cut off some flowering branches while the plant is in bloom, then plop their bases into a bucket of water set near your blooming plant. The bouquet's pollen will remain viable long enough to pollinate your plant. Another alternative is to graft a single branch of a pollinator plant onto your plant. If the pollinator also produces good fruit, then you will be able to harvest two different varieties from your single plant.

Not all fruits require cross-pollination in order to produce a crop. Fruits such as strawberries and raspberries have perfect flowers that are self-fruitful, so you can plant just one variety and harvest a full crop.

Self-fruitful and self-unfruitful are two ends of a spectrum, and some plants lie between these two extremes in their pollination needs. In such cases, it pays to provide for cross-pollination because it induces partially self-fruitful plants to yield more and bigger fruits.

A few plants set fruit without any pollination whatsoever. Among these plants are certain varieties of persimmon, fig, and mulberry. In such cases, not only can you forego providing a pollinator branch or plant, but the fruits themselves are seedless.

## SPACING AND PLANT SIZE

Every fruit plant needs adequate elbow room. A strawberry plant needs its one square foot of space, and a full-size apple tree needs its 500 square feet of space, in each case to permit the leaves of the plant to bathe in sunlight and drying breezes. Such conditions promote good yields and limit disease problems.

The distance that you should set your plant away from other plants, buildings, or walls depends on how big your plant will eventually grow. The eventual size of any plant is determined by the richness of the soil, pruning, and a plant's inherent vigor. Expect less growth of any plant in a poor, gravely soil than in a rich clay loam, unless you fertilize and water diligently. And by pruning, you can keep a plant to almost any size. For example, cordon apple trees—just single stems of fruits—can be planted as close as two feet apart in a row.

Except in cases of extreme soil conditions or careful and constant pruning on your part, the inherent vigor of your plant will be the main determinant of its ultimate size. For many types of fruit plants, you can choose the eventual size of plant you want from a range of sizes. With tree fruits, a full-size plant is called a standard. Dwarf trees are smaller.

A tree might be dwarf because it is naturally small or because it is grafted onto a special dwarfing rootstock. A 'Northblue' blueberry bush is naturally small and will never grow to the size of, for example, a naturally large 'Bluecrop' blueberry bush. A 'McIntosh' apple, however, might grow to a twenty-five-foot tree on one rootstock (MM 111), a fifteen-foot tree on another rootstock (MM 106), or only a six-foot tree on yet another rootstock (M 27). Note that the letters and numbers of the rootstocks refer to when and where they were selected, not the size tree they produce. No matter what the tree size, the fruits will be identical. (But the 'Northblue' fruits are different from the 'Bluecrop' fruits.)

One advantage of dwarf over full-size trees is that they are easier to manage. Most or all pruning, thinning, and harvesting can be done with both of your feet planted firmly on the ground. Smaller plants are also easier to spray, should this be necessary.

Another advantage of smaller plants is that you can cram more of them into a given area. Instead of six bushels of fruit from one large 'McIntosh' tree, you could harvest a couple of bushels of 'Spigold,' a couple of bushels of 'Crispin,' a couple of bushels of 'Golden Delicious,' and a couple of bushels of 'Gravenstein' apples from three dwarf trees occupying the same space as a single large tree. Not only do you get more variety in apples, but because small trees use sunlight more efficiently than do large trees, you also harvest more apples. Or you could plant a couple of dwarf apples, dwarf peaches, and dwarf plums in that same space. With more and smaller trees, you harvest a greater variety of fruits and spread out the harvest over a longer season.

That said, a full-size tree might better suit your needs. Perhaps you enjoy putting up a mess of applesauce or canned peaches all at once. Generally, large trees also tolerate drought, poor fertility, and other adverse soil conditions better than dwarf trees do. With age, a large tree also develops a majestic quality, as well as providing shade and limbs upon which to climb. And besides, for some types of fruits, you have no choice in tree size.

The accompanying chart gives approximate spacing for fruit plants. With these spacings, the branch tips of adjacent plants will just touch, so use this spacing within rows. Allow more spacing between rows so that you can walk or mow down the aisles. Only half the distance listed is needed when planting adjacent to a wall or fence. Bush fruits such as blueberry and pomegranate make attractive, edible hedges, in which case you can set plants close enough to make a continuous row of plants.

| | Approximate Spacing Between Plants (feet) | | Approximate Spacing Between Plants (feet) |
|---|---|---|---|
| Apple | | Jujube | 15 |
| *Dwarf* | 7 | Juneberry | |
| *Semi-dwarf* | 12 | *Bushes* | 6 |
| *Standard* | 20 | *Trees* | 15 |
| Apricot | 15 | Kiwifruit | 8 |
| Blackberry | | Medlar | 10 |
| *Trailing* | 10 | Mulberry | 20 |
| *Erect, semi-erect* | 5 | Nectarine | 15 |
| Blueberry | | Pawpaw | 15 |
| *Highbush* | 5 | Peach | 15 |
| *Lowbush* | 2 | Pear | |
| *Rabbiteye* | 8 | *Dwarf* | 8 |
| Cherry | | *Standard* | 15 |
| *Sweet* | 25 | Persimmon | 15 |
| *Tart* | 10 | Plum | 10 |
| Currant | 6 | Pomegranate | 15 |
| Elderberry | 8 | Quince | 10 |
| Fig | 10 | Raspberry | 2 |
| Gooseberry | 5 | Strawberry | 1 |
| Grape | 8 | | |

# Edible Landscaping

Do not overlook the beauty of fruit trees, vines, and shrubs. The blossoms on a pomegranate or a peach, for example, rival those of plants that are grown only for their flowers. Even the leaves of fruiting plants have a diversity of colors, sheens, and shapes. The rich, forest green leaves of a black currant bush are a perfect backdrop against which to highlight the white petals of daisies. The glossy green leaves of jujube make a nice contrast to the dull green leaves of juniper.

And don't overlook the beauty of some of the fruits themselves: clusters of red currants dangling from the bushes like sparkling red jewels; yellow quinces stuck to the branches like large, woolly spheres; bright red persimmons festooning bare branches of a tree in winter like Christmas ornaments.

Perhaps you want to utilize fruiting plants in the landscape in more fanciful ways—trained into geometric shapes known as espaliers, for example.

*Edible hedge of Nanking cherry, black currant, and saskatoon*

*(continues)*

*(continued)*

Espaliers require detailed and repeated summer and winter pruning but reward you with artistic trees that yield prodigious quantities of high-quality fruit.

*Nanking cherry in bloom*

*Pear espalier at the Cloisters, New York City*

Plan ahead when incorporating fruiting plants into your landscape. Consider a plant's soil and light requirements. Most fruiting plants require well-drained soil (as do most cultivated plants), and if a plant requires full sun, make sure you site it where at least six hours of sun fall each day. Fruit trees such as mulberry and some persimmons drop ripe fruits, so keep them away from patios, walkways, and driveways. If a particular fruit is going to need pesticide sprays, keep that plant far from patios, decks, and other edible plants such as vegetables.

Finally, consider seasonal interest. A plum tree, for example, is at its best (for viewing, not eating) in spring; some varieties can provide triple delight at that time of year, not only for the look of their showy

*An ornamental allee of fruit trees in spring (left), summer (center), and winter (right)*

blossoms but also for their wonderful fragrance. Later, as the blossoms fall, the petals decorate the ground—perhaps a stone path or closely clipped grass—with a temporary delicate, snowy carpet. Blueberry bushes are attractive year round but are most striking in autumn as their leaves turn a crimson color that rivals that of the burning bush, *Euonymous alata*. Just think: with all this, you also get fruit!

Among fruiting plants you will find those suited to almost every landscape use, whether you need a specimen tree in your lawn, a vine to clamber over a pergola, a hedge to define your property, or a groundcover. That fruiting plants lend themselves to such uses should not come as a complete surprise—some, such as kiwifruit, were originally introduced as ornamental plants, their tasty fruits overlooked.

*Hardy kiwi arbor at Mohonk, New York*

# YIELDS

Part of the skill in raising fruit is to produce enough of it in steady supply and over a long season. You will want your plants to yield enough to satisfy your needs, but not so much that the excess is left rotting on the plant or on the ground. (Neighbors and friends can help you out of the latter dilemma.) A planting that bears over a long season lets you spread out the harvest and preserve the bounty at a more leisurely pace and extends the season during which you can enjoy fresh fruit.

The accompanying chart will help you to plan what and how much to grow by giving you the yields for mature plants of each type of fruit. These yields are approximate and will vary with your particular climate as well as the particular varieties you plant. Weather, pests, and the previous year's crops will cause the yields of a particular plant to vary from year to year. This is another good reason to plant a variety of fruits—no matter what the conditions are any season, it will be a good year for some of the plants.

| | Yield at Maturity (lbs.) | | Yield at Maturity (lbs.) |
|---|---|---|---|
| Apple | | Jujube | 100 |
| *Dwarf* | 60 | Juneberry | 20 |
| *Standard* | 300 | Kiwifruit | 150 |
| Apricot | 150 | Medlar | 50 |
| Blackberry | 3 | Mulberry | 300 |
| Blueberry | | Nectarine | 100 |
| *Highbush* | 7 | Pawpaw | 50 |
| *Lowbush* | 1 | Peach | 150 |
| *Rabbiteye* | 15 | Pear | |
| Cherry | | *Dwarf* | 60 |
| *Sweet* | 300 | *Standard* | 300 |
| *Tart* | 100 | Persimmon | 200 |
| Currant | 8 | Plum | 75 |
| Elderberry | 10 | Pomegranate | 100 |
| Fig | 30 | Quince | 50 |
| Gooseberry | 8 | Raspberry | 3 |
| Grape | 15 | Strawberry | 2 |

# Chapter 2

# PLANTING AND GROWING GUIDE

## *S*OIL

*Don't call it dirt. Soil is a more fitting term for this substance that holds your fruit plants upright, supplies them with nutrients and water, and provides an otherwise friendly environment for roots. Soil is actually a mix of things: rock minerals, air, water, and organic matter* (the remains of plants and animals in various states of decomposition). In the ideal soil, about half the volume consists of pores for air and water, and the other half is mostly rock minerals. Organic matter makes up only a small percentage of the soil volume.

The size of the mineral particles influences the size of the pores within the soil. Clay soils are composed of very small particles, so the pores between these particles are also small. Small pores cling to water by capillary action, sometimes so tenaciously that there is insufficient space left for air. (And roots do need air in order to function.) On the other hand, clay soils usually are fertile soils, and the small particles are able to temporarily cling to nutrients that would otherwise wash out of the soil.

Larger than clay particles are so-called silt particles, and larger still are sands. Sands have relatively large pores between the particles and so are well aerated but tend to dry out quickly. Nutrients readily wash through sandy soils, which are relatively infertile anyway.

You can tell what kind of soil you have by feeling it. Squeeze a handful when it is dry and note how well it holds together. Then wet it and try to press it out between your thumb and forefinger into a "ribbon." Rub the wet soil between your fingers: How does it feel?

|       | **Dry**                     | **Wet**                  |
| ----- | --------------------------- | ------------------------ |
| Sand  | Completely falls apart      | No ribbon, gritty        |
| Silt  | Crumbles into various sizes | Short ribbon, silky feel |
| Clay  | Hard to crumble             | Long ribbon, sticky      |

The ideal soil is composed of a congenial mix of clay, silt, and sand particles, giving it a range in pore sizes, some to hold water and some to hold air. Such a soil is called loam.

Most fruit plants require a well-aerated soil, and your first concern with your soil should be that it contains sufficient air. If the soil is boggy because of a high water table, either choose another site, raise plant roots by building up a mound of soil and then planting atop the mound, or lower the water table by digging trenches or by burying perforated plastic pipe to carry water away.

Although organic matter makes up only a small fraction of a soil, it plays an important role. In clay soils organic matter increases aeration; in sandy soils organic matter increases the ability to hold onto both water and nutrients. During decomposition, organic materials release not only nutrients to feed your plants but also other substances that make it easier for plants to take up nutrients already in the soil. Organic materials also feed beneficial soil microorganisms, which help protect plant roots from pests.

Add organic matter to your soil in the form of compost, manure, peat moss, leaves, grass clippings, straw, rotted sawdust, and the like. Either mix these materials into the soil before you plant, or just lay them on top as mulch. Wait at least a couple of weeks before planting if you dig undecomposed materials, such as grass clippings or manure, into the soil. A thick organic mulch laid down after you plant will continue to enrich the soil with organic matter slowly as the mulch decomposes, as well as conserve water and keep weeds in check.

Even though organic materials and the rock particles that make up the soil contain plant nutrients, your plants may need additional nutrients in the form of fertilizer. Before you add fertilizer, however, check your soil's acidity (pH level). In order for a plant to utilize nutrients, the soil's acidity must be in the correct range. Determine your soil's acidity either with a purchased kit or by sending a sample to a testing laboratory. Make your soil more acidic with sulfur or aluminum sulfate, less acidic with ground limestone. The amount of material needed to change the acidity will depend on how much of a change is needed and how much clay is in the soil. To change a soil one pH unit requires, per ten square feet, two-tenths to one pound of limestone, one-tenth to one pound of sulfur, or one-half to six pounds of aluminum sulfate. The lower values apply to sandy soils, and the higher values apply to clay soils.

Periodic fertilization is needed to replace nutrients lost from the soil. Nitrogen is the nutrient needed in greatest quantity. As a rule of thumb, spread about a cup of any fertilizer having about 10 percent nitrogen (such as 10-10-10 or soybean meal) over every 25 square feet of spread of each plant. But keep an eye on plant growth to make sure you are neither under- nor overfertilizing. Be careful, though: growth is a function not only of how much fertilizer you add but also how you prune and how big a crop the plant is carrying.

Also keep an eye on leaf color because deficiencies in nitrogen and other nutrients often cause telltale symptoms. (Again, be careful: diseases, insects, or waterlogged soil can also cause changes in leaf color.) The typical symptoms of nutrient deficiencies include the following:

| | |
|---|---|
| Poor growth and yellowing of oldest leaves | Nitrogen deficiency |
| Yellowing, then browning of leaf margins | Potassium deficiency |
| Yellowing of new leaves, veins green | Iron deficiency |
| Death of growing tips | Calcium deficiency |
| Leaves dwarfed and close together | Zinc deficiency |
| Yellowing of old leaves, veins green | Magnesium deficiency |

If you note any of these symptoms and can truly implicate a specific nutrient deficiency, you can purchase the specific nutrients needed to apply to the soil or, in the case of emergencies, to spray on the plant. It is best, however, to have the soil tested, especially for micro-nutrient deficiencies, because adding too much can be toxic.

If you take care of soil aeration and watering and enrich your soil with abundant amounts of organic matter, you very likely will never see any of these symptoms.

## PLANTING

While balmy weather in late winter or early spring may urge you to plant, other seasons of the year may also be suitable for planting. Fruit plants, like other nursery stock, are sold either with soil around their roots (balled-and-burlapped or in pots) or bare-root.

Plant balled-and-burlapped or potted plants anytime of the year that the ground is not frozen. Plant bare-root plants only while the plants are still dormant and leafless.

You may want to reserve that primal urge to plant in spring for annual flowers and vegetables, because in many ways fall is a better season in which to plant fruits. In fall the soil usually is neither sodden nor bone dry, but moist and crumbly—just right for digging holes. And remember that the balmy weather of spring is also going to lure you to pruning, readying the vegetable garden, and mowing the lawn. Planting fruits in the fall leaves you with one less thing to do in the flurry of spring's activities.

Even the fruit plants themselves can benefit from fall planting. Soil temperatures cool slowly in fall, allowing some root growth even on leafless plants. And when spring finally does arrive, the plant is already in place, watered through the winter and ready to grow. Having the plant already in place is especially important with plants such as gooseberries and currants, which leaf out very early in spring. Just a few plants—peach, plum, apricot, cherry, persimmon, and pawpaw—do not take kindly to bare-root planting in the fall.

When you receive plants from a nursery, whether or not they are bare-root, make sure their roots are moist. If you cannot plant bare-root plants right away, keep them cool to hold back growth. After you check that the roots are moist, wrap the roots of small plants such as strawberries or brambles in a plastic bag and store them in your refrigerator until you are ready to plant. Hold trees, shrubs, and vines until you are ready for them by temporarily planting them in shallow holes in a shaded location.

Just before you are ready to plant, soak the roots of bare-root plants in water for a few hours. After the soaking, trim back the roots, shortening any lanky roots to about eighteen inches and cutting back any diseased or frayed roots to healthy portions. With strawberries, use scissors to cut all the roots in a bundle of plants back to four inches. While you plant, keep all roots in water or covered with moist burlap to prevent them from drying out.

Make sure the soil is neither too wet nor too dry before you begin digging, because either extreme can make work frustrating for you and can ruin the pore structure of the soil. The soil is moist and ready for digging when a handful easily crumbles as you gently squeeze it in your hand. If you are planting in lawn, use a flat-bladed shovel first to skim off the top inch of soil wherever you are going to plant. (Add the sod to your compost pile, or use it to patch other areas of your lawn.)

As you dig planting holes, mix in certain soil amendments that would otherwise only work slowly down to plant roots. If the pH needs adjustment, this is the time to raise the pH with limestone or to lower it with sulfur or aluminum sulfate. Phosphorus is a nutrient that moves slowly down through the soil, so if a soil test indicates that phosphorus fertilizer is needed, also mix this in with the soil of the planting hole. Also add some well-rotted humus, such as compost, peat moss, or aged sawdust, to increase both the aeration and the water-holding capacity of the soil. Do not add too much of any organic material, because if the soil within the planting hole is too loose and fluffy as compared with the surrounding soil, roots have little inducement to grow into surrounding soil.

*Setting tree in planting hole*

Beyond the planting hole, spread lime, sulfur, phosphorous fertilizer, or organic materials on the surface of the ground. By the time roots spread, these materials will have percolated down into the soil.

Dig a planting hole only large enough to accommodate a plant's roots. After you dig the hole, roughen up its sides and bottom with your shovel so that the roots have an easier time penetrating into the surrounding soil. Then put some soil back into the hole, forming a mound on which to set your plant. Adjust the height of the mound so that the soil line on the stem will be about the same height as it was in the nursery, allowing for a slight amount of settling. For strawberries the soil line should go right across the middle of the crown so that the tip of the crown neither suffocates beneath the soil nor dries out from being too high out of the soil. Splay out the roots of bare-root plants over the mound or, in the case of balled-and-burlapped or potted plants, comb out the roots on the outside of the root ball to get them growing outward.

Backfill the soil from the planting hole a little at a time, tamping it among the roots with your fingers or a stick. Occasionally bounce bare-root plants up and down slightly as you fill the hole to help work soil among the roots. When you finish filling the hole for larger trees, vines, or shrubs, form a rim of soil two to three inches high and two to three feet in diameter to act as a catch basin for water.

Cover the bare soil with two to three inches of organic mulch to help water percolate into the soil, prevent evaporation of water from the surface, and suppress weeds. Mulch is especially important following fall planting in locations where winters are cold, to keep the soil warm as long as possible and to prevent heaving of new plants, not yet firmly anchored, as the soil alternately freezes and thaws. Keep mulch a couple of inches away from the trunks of trees to discourage crown rot and rodent feeding.

Now is the time to put the stake in the ground for newly planted trees or vines needing permanent or temporary staking. Drive a metal post or a sturdy, rot-resistant wooden post into the ground a few inches away from the trunk, on the side from which the strongest winds blow. Fasten the trunk to the stake with soft rope or old inner tubes, first tying the material tightly around the stake, then loosely around the plant.

See the "Pruning" section in this chapter (page 22) for any pruning necessary immediately after planting.

Finally, water the plant to get it off to a good start and to settle the soil. Slowly pour a couple of gallons

# Plant Propagation

Plants are propagated either by sowing seeds or by cloning.

Cloning, or using a piece of a plant to make a whole new plant, is the method used with most fruit plants. The following are traditional ways of cloning plants.

*Stem cuttings (of grapes, for example):* Stems stuck into the ground grow roots and new shoots.

*Root cuttings (of blackberries, for example):* Short pieces of root placed in soil grow shoots and new roots.

*Layering (of gooseberries, for example):* Stems are bent and held down to form roots where they contact soil; the rooted stems are cut off and transplanted.

*Grafting (of apples, for example):* A stem from the plant to be propagated is cut off and joined to a rootstock a few inches above the soil. The stem section gives rise to all growth above the graft.

In each case, the whole plant (except, in the case of grafted plants, for the roots and a few inches of trunk below the graft) is exactly the same as the original plant.

When you propagate plants by sowing seeds, however, the resulting seedlings are different from each other and from the parent plant. In the case of a few fruits—alpine strawberries and, to a lesser extent, peaches—differences are only slight. But with most fruits, the differences are dramatic. If you were to take seeds from a 'McIntosh' apple and plant them, the resulting seedlings would eventually bear fruits quite different from each other and from 'McIntosh' apples.

Because seedling plants are different from their parents, seedlings are used to develop new varieties of a fruit. A seedling that is superior will be given a name, then propagated by cloning. 'McIntosh,' 'Bartlett,' and 'Concord' are examples of variety names (sometimes called cultivar names) given to worthy apple, pear, and grape seedlings. Such seedlings may have been deliberately sown or discovered in the wild.

Besides being of predictable quality, a clonally propagated fruit plant also bears its first crop sooner than would a seedling. A seedling apple tree, for example, might not yield any fruit for a decade or more, but a grafted apple could bear a few fruits two seasons after you set it in the ground. And even after all those years, you are taking a chance with the fruit from a seedling apple. On the average, only one in 10,000 apple seedlings will bear fruit that tastes as good or better than the parent.

within the catch basin beneath trees, proportionately less for smaller plants.

Do not turn your back on the plant for the season. Your careful planting will be for naught unless you diligently weed and water during the plant's first—and most critical—year in the ground.

# WATERING

A plant needs water to keep cool, to pump minerals up to its leaves, and to form sugars. Under certain conditions, a plant may be able to fend for itself on rainfall, but there are times when a plant needs your help—in the form of irrigation.

Before you even touch your hose spigot, there are a few things you can do to help plants make maximum use of natural rainfall and irrigation. Maintain high levels of organic matter in the soil to help it hold water. Cover the ground with mulch to slow evaporation from the surface. If the mulch is a fluffy, organic material such as straw or leaves, it also will keep the surface loose so that water can penetrate rather than run off. Weeds pull water from the soil, so control them to leave more water available to your fruit plants. Finally, contour the surface of the ground with low mounds to catch and hold water.

Water needs vary with the soil type, the climate, the season, and the particular plant. Sandy soils need most frequent watering. Low humidity, wind, and heat all increase the water requirements of plants. And plants drink the most water when they are growing most vigorously.

Large plants use more water than do small plants, but large trees may be better able to fend for themselves because their roots reach deep into the soil for water. Berries and most dwarf trees have shallow roots and may or may not need supplemental watering throughout their lives, depending on natural rainfall. Newly planted trees, shrubs, and vines are not self-sufficient. Get them off to a good start with regular watering during their first season in the ground.

A reliable way to tell whether any plant needs water is to dig a hole and feel the soil for moisture. An alternative method, one that avoids pocking your soil with test holes, is to use an inexpensive electronic soil probe. If your fruit plant tells you it is dry by wilting, water it immediately. Once wilting proceeds beyond a certain point, the plant will not recover.

As a rough estimate, give berry plants and small fruit trees one inch of water per week during active growth. Larger trees may need double or triple this amount. A rain gauge or any straight-sided container can tell you when an inch of rain has fallen. If sufficient rain has not fallen, irrigate to make up the difference. That inch depth of water is equivalent to about a half-gallon of water per square foot. To estimate the gallons (rather than the depth) of water a plant needs, estimate the amount of square feet covered by the roots—which is approximately the same as the spread of the crown—and apply that many gallons of water.

There are two exceptions to the one-inch-per-week (or half-gallon-per-square-foot) rule. The first exception is for fruit plants in containers. Such plants may need water every day—perhaps even twice a day—during their peak of growth in summer. Check the soil for dryness, then apply water until it flows out the drainage holes in the bottom of the container. The second exception is toward the end of the growing

## Drip Irrigation

Decades ago an agricultural scientist in rain-parched Israel took note of the luxuriant growth of plants (weeds) beneath a leaky spigot and decided to try this system—deliberately—on crop plants. So-called drip, or trickle, irrigation makes the most efficient use of water by slowly and almost continuously replacing water removed from the soil by plants. With drip irrigation the soil does not cycle between periods of excessive wetness and dryness, as is the case with watering either by flooding or by sprinkling. Drip irrigation also pinpoints water application, so water is not wasted on paths or between your fruit plants, where it can promote weed growth. The leaves of drip-irrigated plants are not wetted, lessening the chance of disease.

The bare bones of a drip-irrigation system consists of a header—a tube, usually half-inch black plastic pipe—that brings water from the hose spigot to various points in your garden, with emitters plugged into the header. Emitters typically drip one-half to one gallon of water per hour, and you might allow emitters to run for a half-hour to an hour per day. For a continuous row of plants, such as strawberries or hedgerow raspberries, emitters need to be spaced as close as a foot apart in order to wet a continuous strip. Individual trees or bushes each need one or more emitters, depending on the size of the plants.

*Drip-irrigation spot emitter near blueberry bush*

The emitters themselves are technological marvels. In the early days of drip irrigation, emitters were plagued with problems: they clogged easily, and their water output varied with the water pressure and the location of the emitters along the lines. Today you can purchase emitters that are

*(continues)*

*(continued)*

self-cleaning and pressure-compensating. There are also microsprinklers—emitters that are miniature sprinklers—that spread the water near ground level out over small areas.

Drip irrigation is as friendly to gardeners as it is to plants. A system requires neither high water pressure nor a large investment of money. Perhaps best of all, drip irrigation is easily automated.

A typical drip-irrigation setup consists of the following components, beginning at the hose spigot: an anti-siphon valve, to prevent water from siphoning back into the water supply in case of a drop in water pressure; an automatic timer, which turns the water on and off one or more times each day; a pressure regulator, to decrease the pressure in the line (to about 10 psi, depending on components) and even out variations in water pressure; a 150- to 200-mesh filter, to screen out particles in the water; header pipe and emitters, easily and inexpensively fitted together because of the low pressure of the system.

Drip irrigation, like any watering system, is not foolproof. No dramatic columns of water spurt into the air to tell you everything is working, so you must periodically give a drip system a close inspection.

*Drip-irrigation components near spigot*

season, when all plants need to slow down growth and harden off for winter. This is the time to gradually cut back on watering.

The three ways to water a plant are by flooding the ground, by sprinkling, or by drip irrigation. For greatest efficiency when flooding the ground around a group of plants or a single plant, build up the soil to form a catch basin to contain the water. And when you turn on the hose, do not let the water flow out so fast that it washes away surface soil or mulch. The potential for eroding the soil and the inconvenience of having to move the hose from plant to plant limit the usefulness of flooding as a method of watering.

Sprinkling is an easy way to water, especially if you have sprinklers set up at permanent locations. The disadvantages of sprinkling are that the wet leaves are prone to disease and that the diffused spray will also water weeds trying to grow between your fruit plants. The ideal time to sprinkle is in the morning, early enough so that airborne droplets are not blown by wind or evaporated in hot sun, yet late enough so that the leaves dry quickly after you turn off the sprinkler.

Drip irrigation is a recent innovation in watering plants. Because it is easily automated, drip irrigation is also ideal for watering fruit trees in containers. For more information about this method of irrigation, see the accompanying box (page 20).

No matter what method of watering you use, water regularly whenever rainfall is insufficient. Besides promoting good growth, regular watering helps prevent nearly ripe fruit from cracking—a particular problem with fruits such as prune plums and 'Winesap' apples.

On the other hand, do not be overzealous with your watering. Overwatering is wasteful of water and, by suffocating roots, is as harmful to a fruit plant as is underwatering.

# PRUNING

The reasons to prune any fruit plant vary and depend on the kind of plant and whether it is young or old. In any case, to produce good yields of tasty fruits on manageable plants, all fruit plants need some pruning. Most need annual pruning.

The tool that you need for pruning depends on the thickness of the stem you intend to cut. Your thumbnail is all you need to pinch out the succulent growing tip of a blackberry cane in summer. Use a pair of pruning shears to cut stems up to about a half inch in diameter. A long-handled lopper will cut through stems an inch in diameter, perhaps two inches, depending on the design of the lopper. For larger limbs, use a pruning saw. The teeth of these saws—different from those in woodworking saws—are designed not to become clogged by wet sawdust. For stems out of reach, use a pole pruner, which has a pruning saw and/or lopper at its end.

*Pruning tools*

When you prune a stem, you either shorten it or remove it completely. Shortening a stem is called heading, and this type of cut causes buds along the remaining portion of the stem to grow. The more vigorous the stem and the more that you cut off, the more vigorous the regrowth. (This invigoration is localized on the stem; in terms of the whole plant, any pruning has a stunting effect.) Use heading cuts where you want branching to occur and where you want to stimulate growth that was weak.

In contrast to heading, complete removal of a stem is called thinning. Usually no regrowth near the cut results following thinning, so this type of cut is useful when you want to open up a dense portion of a plant to light and air. Thinning cuts also are useful when you want to keep a plant from growing any bigger than it is.

*Limb removal to collar*

Pruning wounds a plant, so your cuts should be made in a way that promotes rapid and complete healing. Because clean cuts heal best, keep all pruning tools sharp. Most pruning of fruit plants is done when the plants are dormant and leafless. The ideal time to prune during this dormant period is in late winter or early spring, just before growth begins, because this is when wounds heal most quickly.

For best healing when heading a stem, cut just beyond and sloping away from a bud. If you cut too close to a bud, it will dry out and die. If you cut too far from a bud, a stub will be left, which will die and provide possible entrance for disease.

When you make a thinning cut, remove the stem or limb just back to the collar at its origin. If the limb is large, do not remove it with a single cut or bark may strip from the trunk as the limb falls. Instead, first cut back the limb so that it is only about a foot long, beginning with an undercut before you saw the limb off from above. With the bulk of the limb removed, support the remaining portion with your hand as you saw it off at its collar.

## Pruning a Young Tree

Direct the growth of a young tree to develop sturdy and well-placed limbs that are open to sun and breezes and capable of supporting loads of fruit. Pruning is only one of the techniques used in training a young tree. Do not prune a young tree any more than is absolutely necessary, or else you will unduly delay the time until first harvest.

Begin pruning your tree just as soon as you set it in the ground. If the plant is just a single unbranched stem, shorten that stem to about thirty inches above

*Open-center tart cherry showing limbs crowding with age*

*Top: Toothpick spreading branches*
*Bottom: Clothespin spreading branches*

ground level. Cutting it back will induce branches, which will become permanent limbs, to grow. The location of these branches will remain the same throughout the life of the tree, so as the branches develop, select carefully those you want to save and cut away all others as you notice them. The lowest branch should be no less than two feet above the ground, with subsequent ones arranged in a spiral fashion and six inches apart up the trunk. Well-anchored branches are those that are narrower than the trunk and attached to it at wide angles. If a well-placed branch does not come off the trunk at a wide angle, spread it by inserting a clothespin or a toothpick between a branch and the trunk. Where a well-placed branch is too thick in relation to the trunk, weaken the branch by pinching its growing point or cutting it back.

Your new tree may already have branches. In this case, select three or four healthy and well-positioned branches to become permanent limbs, then cut away all others. Induce vigorous, spreading growth from the branches that you save by shortening each so that it ends in an outward-pointing bud and is only a few inches long.

Fruit trees are commonly trained to one of three forms: central-leader, modified-central-leader, or open-center. (See the accompanying box for descriptions of these forms.) If you are going to train your tree to a central-leader or modified-central-leader form, allow the topmost bud to grow upright as a continuation of the trunk, called the leader. For an open-center tree, allow three or four branches growing outward and upward to become main branches, then cut the trunk back to just above the topmost branch so that there is no continuation of the trunk beyond this point.

Some pruning is needed in subsequent years of training your young tree. For a central-leader tree, cut back about a third of the previous year's growth of the leader each year to stimulate additional tiers of branches. Select branches to become permanent limbs just as you did the first season, removing all others. When the trunk on a modified-central-leader tree is about eight feet high, shorten it to a weak side branch to discontinue extension of the leader. For the open-center tree, head back the three or four branches growing off the trunk to get them to branch further.

The only other pruning required on a young tree is to remove suckers and water sprouts. Suckers are vigorous shoots originating at or near ground level. Because they originate from the rootstock of grafted

*Top: Sweet cherry being trained as central-leader*
*Bottom: Suckers at the base of an apple tree*

## Pruning a Mature, Bearing Tree

In the case of a mature tree—one that has reached its full size and is bearing regular crops—use your pruning tools to keep the tree healthy, productive, and within bounds.

First look over your tree for branches that are diseased or have the potential to become diseased. Dark, sunken areas or little black or red specks on the bark are evidence of disease. Cut any diseased branches at least six inches back into healthy wood. Also cut away any dead or broken branches, both of which can provide entryway for disease. Early in the season you will be able to recognize dead branches because their buds remain lifeless while those on healthy branches are swelling.

Next, go over the tree with thinning cuts. With age, it is common for limbs high in a tree to overgrow the lower ones. Counteract this tendency—which would result in the tree canopy becoming too tall and shaded—by totally removing some of the upper limbs. Dense growth shades the tree, decreasing productivity and, along with poor air circulation, increasing the risk of disease. Wherever growth is too dense, use thinning cuts to open up space within the tree. Also cut away water sprouts, suckers, and crossing branches and generally open up the tree where growth is too dense.

Any tree will also have a certain amount of twiggy, nonproductive wood. Thin some of this out and head some back where you want to force the remaining buds into vigorous growth.

*Pear water sprouts*

trees, they would bear undesirable fruits if allowed to mature. They would also ruin the form of the tree. Water sprouts are vigorous, upright shoots that develop along the branches and are undesirable because they shade the tree and produce little or poor-quality fruit. Cut them away or snap them off as soon as you notice them.

Finally, most fruit trees require a certain amount of heading cuts just to stimulate the growth of young wood to replace old wood. The amount of stimulation needed varies, depending on the fruit-bearing habit of the tree. Peaches, for example, bear fruit only on wood that is one year old and therefore need to make enough new growth each year for the following year's crop. Apples and pears, on the other hand, need less pruning because they bear fruit on old wood.

Do not bemoan the loss of fruit buds as you prune your tree. In removing some potential fruits before they form, you enable the tree to channel more energy into the fruits that do form, making them bigger and better-tasting.

## Pruning a Neglected Old Tree

The problem with a neglected old tree is that it has grown too tall and dense. The tree shades itself to the extent that the fruits—if any—are borne only high up in the tree. A neglected old tree also is commonly over-burdened with dead and diseased wood.

Begin by lowering the tree, using your pruning saw to cut back major limbs. Do not do this in one season, or the once-shaded bark will sun scald. Depending on the severity of pruning needed, you may have to spread these major cuts out over three or four years. Also cut away any dead or diseased wood.

Vigorous water sprouts will grow from large pruning cuts. Cut away most of these, but save a few to shade the bark and as replacement limbs for your now shorter tree. Select new limbs and branches similarly as you would for a young tree.

### Tree Forms

For sturdy limbs bathed in light and air, fruit trees are most commonly trained to one of three forms: central-leader, open-center, or modified-central-leader. The central-leader tree is shaped like a Christmas tree, with an upright trunk off which grow permanent scaffold limbs, increasing in length from the top to the bottom of the tree. The open-center tree has a short trunk capped with three to five scaffold limbs radiating outward and upward, giving the tree the shape of a vase. The modified-central-leader is a combination of the other two forms. When the trunk of a tree being trained to a modified-central-leader reaches a height of about eight feet, cut it back to a weak branch. This stops further upward growth of the tree.

Fruit trees differ in their natural growth habits. If a plant is naturally inclined to grow in one of the three forms, coax it in that direction rather than trying to force it into another.

# Fruit Thinning

In order to get through the critical period just following bloom, when late frost might knock off some developing fruits, most fruit trees initially set more fruits than they can ripen. Once the danger of frost is well past, a month or two after bloom, trees naturally shed excess fruits. Because this shedding occurs in June over much of the country, it is called June drop and should be no cause for alarm.

A tree just wants to ripen fruits; we gardeners, however, want the tastiest and often the largest fruits. To harvest such fruits, further thinning—removal of fruits—is needed. Fruit thinning is also a way to reduce a crop that otherwise might break off a limb and to prevent alternate bearing, a tendency of some fruit trees to alternate years of very large and small crops.

*Hand-thinning apple fruitlets*

*Fruit set on apple*

For maximum benefit to remaining fruits and to next year's crop, thin fruits as soon as they set, right after bloom. For insurance, however, leave some extra and thin again after the June drop. When you thin fruits, selectively remove those that are smallest or injured by insects or disease. The spacing of the remaining fruits varies with the type of fruit and is spelled out in the plant portraits for each fruit. There is no need to thin small fruits (such as cherries) or bush fruits (such as blueberries) beyond the fruit bud removal that occurs as you annually prune these plants.

## General Directions for Pruning Bushes

Bushes, in contrast to trees, do not have single trunks. Instead, they annually grow many vigorous new stems from near or below ground level. The distinction between a tree and a bush is not always clear-cut. Fig, pomegranate, and quince, for example, can wear either guise. And some definitely bushy plants, such as gooseberry and currant, can, if desired, be forced into becoming trees—small trees but, with their single trunks, trees nonetheless.

Pruning of a newly planted bush is not critical. On bushes that sucker profusely, cut back all the stems to nearly ground level to force growth of new suckers. For bushes that make only a few suckers, do nothing to the branches or, if the plant seems top-heavy, just shorten them slightly.

Once mature, bushes are generally pruned by renewal, which means that over the course of time old wood that is no longer productive is replaced by younger wood. Cut away old wood at or near ground level and thin out young stems so that the bush does not become overcrowded.

How much to leave and how much to prune away depends on the particular type of bush fruit: how much it suckers and the age at which the wood is most fruitful. Black currants represent one extreme, suckering profusely and fruiting mostly on wood that grew the previous season. The way to prune a black currant bush is to thin out all but a half-dozen of these one-year-old stems (for fruit the following season), then to remove, or shorten to a vigorous one-year-old branch, any wood more than a year old.

*Black currant before pruning*

*Black currant after pruning*

Blueberry is an example of the other extreme in bush fruits, suckering only moderately and bearing fruit on wood that is even four years old (actually, on younger branches growing off this wood). Such bushes need less pruning than those that sucker profusely. All that you need to do with these bushes is to cut back stems that are drooping to the ground or too dense in the center of the bush, as well as unfruitful twiggy growth. And then occasionally cut away nonproductive old wood to make way for a young replacement shoot.

You may wonder how strawberry fits in with other bush fruits. Strawberry is actually a bush whose stems are telescoped down, with only a fraction of an inch from leaf to leaf. Prune strawberries by thinning out excess plants, as discussed in the plant portrait for this fruit.

## General Directions for Pruning Vines

All fruiting vines need a trunk off which to grow temporary fruiting arms, or else long, permanent arms, called cordons, which in turn give rise to the fruiting arms. To stimulate vigorous growth of the trunk-to-be, cut back your vine at planting to a single robust shoot, then shorten it to two or three healthy buds. From then on, training and pruning varies with the kind of fruit and is discussed under the heading for that fruit.

# SPECIAL GROWING TECHNIQUES

## Bringing Fruit Trees into Bearing

In any fruit tree a balance exists between shoot growth and fruiting, with more of one being offset by less of the other. Because they have not yet borne fruit, young fruit trees commonly make very vigorous shoot growth. While a certain amount of shoot growth is desirable for any tree to fill its allotted space, shoot growth sometimes persists at the expense of fruiting. Various techniques can slow a tree down and bring it into a bearing habit. (Fruit *bushes* rarely have a problem settling down into a bearing habit.)

The first technique to coax a tree into bearing, branch bending, should be started early in a tree's life. Because upright shoots are inherently more vigorous than are horizontal shoots, merely bending branches downward slows growth and induces fruit buds to form. Bring the branches down to near horizontal with weights, with string secured to the trunk or to stakes in the ground, or with notched wooden branch spreaders. If you want some additional growth on a branch, do not bring it all the way down to the horizontal but leave it at a slight angle. And do not arch the branch as you bend it down, or else a vigorous shoot will grow from the high point in the arch.

*Weight on apple branch*

A second technique to hasten fruiting is bark ringing. During the period between when the petals fall and a month after blossoming, cut a thin ring of bark from the trunk. Make two parallel cuts about $1/4$ inch apart around the trunk, then peel away the strip of bark between them. The bark heals within a few weeks of ringing. Do not use this technique on stone fruits (peach, apricot, cherry, nectarine, and plum), which are too easily infected at wounds.

In any one year, use one—not both—of these techniques on a fruit tree. At the same time, consider what else, besides youth, might be making a tree overly vigorous. Are you overfertilizing? Are you pruning too drastically?

Do not use either of these techniques on weakened trees. Such trees need to be invigorated before they can begin fruiting and may even be killed by bark ringing.

## Fruit Trees in Containers

For centuries gardeners have enjoyed growing fruit trees in containers. An entire wing of the French palace of Versailles was devoted to potted orange trees. Orange trees are not cold-hardy in that part of France, so the plants spent winters indoors in their *orangerie.*

An ornamental fruiting plant in a pot might be just the plant to decorate your sunny terrace or your balcony if you live in an apartment and have no land. Many tropical and subtropical fruit plants, such as fig, natal plum, and, of course, orange, are ideal subjects for container growing in northern areas. Also consider growing decorative cold-hardy fruit plants. An added advantage of growing such plants in pots is that you can move them to shelter if frost threatens their blossoms in spring.

*Potted fig tree*

Just about any fruit plant can be grown in a container, but best suited are those that are dwarf and lack taproots. The plants may be naturally dwarf or made so by being grafted onto a dwarfing rootstock. Most fruit plants do not have taproots, notable exceptions being persimmon and pawpaw.

Any standard potting mix is suitable for fruit plants (use a very acidic mix for blueberries, though), as is any container, as long as it has drainage holes. Except for strawberries, which you can grow even in six-inch pots, use a pot at least a foot-and-a-half wide and deep. Fertilize as you would any other potted plant, slackening off toward the end of summer to allow the plant to harden off for winter.

Repot your plant each year into a larger container as it grows. Once the plant gets as big as you would like it to be (which may be determined by how large a pot you can move around), continue to repot it every year or two, right back into the same pot in which it was growing. Root-prune when you repot to make space for new root growth. Do this by tipping the plant out of its pot and slicing off roots around the edge of the root ball. Then put the plant back into the pot, packing new potting soil into the space between the root ball and edge of the pot. To compensate for each year's loss of roots during repotting, prune the branches more severely than you would for a plant growing in the ground.

Fruit plants in containers have two special needs: attention to water in summer and a suitable home for the winter. Roots in pots cannot explore as much soil as can plants in the ground, so in midsummer's heat, a potted fruit might need watering every day.

If you are going to be away from home for only a day or two, temporarily moving a plant into the shade will relieve you from watering chores, but a more permanent solution is needed for longer absences or if you do not want to be burdened by having to water daily. A drip-irrigation system can automatically water potted plants, using special emitters for this purpose. With a well-drained potting mix, plants will not be harmed by once- or even twice-a-day watering from the time they are growing strongly in spring until they slow down in fall. Water by hand in fall and winter, when overwatering might cause roots to rot.

*Trimming roots off potted tree*

Because the roots of potted plants are not nestled within the earth, they must be given additional protection from winter cold. The most obvious solution is to give them the protection of the earth—by plunging them, pot and all, into a hole in the ground. Or protect

*Combing roots from potted tree*

*Repotting*

plants left sitting on top of the ground by piling a loose, insulating mulch such as wood chips or leaves up and around the pots. A third choice is to move the plants somewhere that is cold but not frigid, such as an unheated garage, shed, or basement. One place

definitely not suitable for cold-hardy plants in winter is inside your home; these plants must have a period of cold weather before they will resume growth in spring.

Once protected for the winter, plants need little further care. One thorough watering may be all the plants need until they resume growth again in spring. Because the plants are leafless in winter, they do not need light.

Tropical and subtropical plants cannot tolerate very cold temperatures and so should spend the winter indoors. Because they retain their leaves in winter, these plants need the bright light of a southern window. Generally, the warmer the indoor temperature, the more light the plant requires. Keep in mind that subtropical fruits, such as oranges, enjoy cool rather than warm indoor temperatures in winter.

With attention to winter quarters, watering, root pruning, and fertilizing, fruits growing in containers obviously require more effort than those growing in the ground. But working closely with the plants and watching them respond can be a reward in itself—along with delectable fruits.

# Chapter 3

# Plant Portraits

*H*ere are details about the fruits you choose to grow or perhaps already are growing. *Each entry, listed alphabetically by the fruit's common name, provides specific information about the best site for the plant, as well as directions on growing the fruit and a listing of major pest problems. The portraits conclude with a "Selection of Plants," which is a listing*

of varieties and, in applicable cases, rootstocks. The variety lists are not meant to be complete—after all, there are more than 5,000 varieties of apples. Instead, the lists are of notable varieties recommended for planting because of their flavor, disease resistance, or other qualities.

**Note:** Some of the photographs refer to fruit varieties that are not specifically discussed in the plant portraits. In these instances, follow the planting and growing instructions for the general variety and you will do fine.

Please use the plant portraits in conjunction with other sections of the book. In chapter 2, for example, information was provided on various ways of pruning a fruit tree. Put that information to use when you start training your peach tree to an open-center form, as recommended in the portrait for peaches. Major pests are listed for each fruit. Recommendations for controlling these pests with pesticides are periodically updated, so read product labels for specific recommendations. Chapters 4 and 5 provide general information on harvesting and pest control.

# PLANT PORTRAIT KEY

**Common Name** of a fruit begins each section and is in boldface type.

*Latin Name* of a fruit is in italic type.

**Pollination needs:** A "self-fruitful" plant can bear fruit in isolation. A plant that needs "cross-pollination" will not set fruit unless there is a different variety of the same fruit growing nearby. A plant that is "partially self-fruitful" sets more and larger fruits when cross-pollinated.

**Sun/shade requirements:** These requirements refer to the average hours of sun needed per day. Full sun plants require six hours or more of strong, direct sunlight per day; partial shade plants require three to six hours of direct sunlight per day; shade plants require two hours or less of direct sunlight per day.

**Zones:** Check "The USDA Plant Hardiness Map" (page 118), which is based on average annual temperatures for each area or zone of the United States, to see what plants are appropriate for your climate. For best results, match plants appropriately to your climate zone.

# APPLE
*(Malus domestica)*

**Pollination needs:**   Requires cross-pollination.

**Sun/shade requirements:**   Full sun

**Zones:**   3 to 9

Apple is the fruit most people consider when they think of growing fruit. And with good reason, because apples can be grown almost everywhere with the choice of suitable varieties. Rootstocks offer a range in tree sizes, from plants that never grow taller than six feet to those towering at forty feet.

Apples are traditionally associated with autumn, but if you are a real fan of apples, grow varieties that ripen in summer, as well as late apples that store well until spring.

## Growing

The size of your property should not keep you from planting apples, because a dwarf tree—which produces full-size fruits—requires only six feet or less of space around it. But choose the site with care. Full sun is best. Apples bloom fairly early in spring, so choose a site that is not prone to late spring frosts.

Apple trees tolerate a wide range of soils, but prefer one that is not waterlogged, especially during the growing season. The soil should be moderately fertile and slightly acidic. If you prune, fertilize, and water correctly, the shoots on a young tree should grow

about three feet in a season, while those on a mature tree should grow about eighteen inches.

Right after you plant an apple tree, begin training it to a central-leader, modified-central-leader, or open-center form. Any of these forms is suitable for apple, although upright varieties such as 'Delicious' tend to grow as central-leaders, while spreading varieties such as 'Golden Delicious' tend to grow as open-center trees.

Be careful about pruning apple trees when they are young. Too much pruning, especially of standard trees, stimulates vigorous growth instead of fruiting. On the other hand, do not allow a dwarf tree to bear too many fruits at too early an age, or growth will be stunted. If that does happen, prune more severely to stimulate growth, then grit your teeth and pull off some young fruits so that the tree can channel its energy into growing shoots.

Once a tree reaches full size and fruits regularly, prune annually to keep the tree open and stimulate some new growth to replace very old wood that is no longer fruitful. Apple trees bear fruits on short, stubby spurs on older branches and therefore do not need much growth stimulation each year for fruit. When spurs become overcrowded or too old, remove some of them and cut away a small branch or two on others.

Pruning branches and spurs removes fruit buds and hence potential fruits—but not enough. Left to itself, an apple tree will overbear in most years. Each flower bud opens to five flowers, only one of which should be allowed to develop into a fruit. And even then, where flower buds are close together, thin fruits further so that they are five inches apart. If all this fruit thinning seems excessive, keep in mind that only five percent of the flowers normally need to set fruit in order to yield a full crop of apples.

## Pest Damage and Identification

A few major insects and diseases attack apple, and one or more of these will require your attention in all areas except perhaps parts of the West Coast.

Worldwide, the most serious disease of apple is apple scab, which produces corky brown lesions on the fruits and leaves. The disease overwinters on old infected leaves on the ground, then continues to spread during the growing season whenever the weather is warm and wet. Besides spraying, control this disease by cleaning up or burying leaves in the fall, pruning so that light and air can rapidly dry leaves and fruit, and growing scab-resistant varieties such as 'Liberty,' 'Golden Delicious,' and 'Crispin.'

Cedar-apple rust disease produces orange spots on leaves and fruits. Because cedar-apple rust needs both cedar and apple trees to complete its life cycle, the disease is especially prevalent in the East, where wild cedars abound. Disease spores travel in the air for miles, so removal of cedar trees is not a practical means of disease control. Instead, plant resistant varieties such as 'McIntosh' and 'Liberty' or else spray.

Fire blight disease causes branches and leaves to look as if they have been singed by fire. Blackened leaves remain attached to branches, and the tips of new shoots curl over like shepherds' crooks. Again, one control is to plant resistant varieties. Also, do not stimulate very succulent growth, which is susceptible to fire blight attack, by heavy pruning or fertilization.

You can control this disease by pruning if you are diligent in removing infected branches. Cut six inches into healthy wood and sterilize the pruning shears by dipping them in alcohol between cuts.

Pruning also helps control so-called summer diseases such as black rot, bitter rot, and white rot, all of which cause soft, rotten areas on fruits. Look over branches as you prune, keeping an eye out for dark, sunken cankers of disease.

Powdery mildew disease coats leaves with a white coating and russets fruit. Control powdery mildew by pruning away dormant buds in winter that show white fungal growth, by planting resistant varieties, or by spraying.

Three insects that may ruin your fruits are the codling moth, the plum curculio, and the apple maggot. If you bite into an apple and find a large white "worm," that is a codling moth larva. You also may notice one large hole on the outside of the fruit, where the larva entered. Plum curculio, present east of the Rockies, may cause a fruit to fall as it leaves a crescent-shaped scar. But if the fruit does not fall, it is perfectly edible, with only a superficial blemish. Apple maggot, also found in the East, tunnels throughout the fruit, dimpling the outside and leaving brown trails within.

Use alternative measures to augment or replace spraying for insect pests. Codling moth larvae prefer to enter fruits that touch each other, so thin the fruits to decrease damage. Pheromone traps are available for codling moth and can be effective with isolated trees. Codling moth and plum curculio are most active early in the season, so you can stop insecticide sprays after

*Plum curculio damage on apple fruit*

*Apple maggot trap*

about six weeks following bloom. Then the apple maggot becomes active, but you can effectively trap this pest on fake apples—red spheres coated with sticky Tangletrap that are hung in the trees. Hang one trap per dwarf tree, four to eight per large tree, at eye level and a couple of feet within the canopy but not obscured by leaves.

## Selection of Plants

For a plant that has been enjoyed for thousands of years, it is no wonder that there are thousands of apple varieties. Not only are there numerous varieties, but also many rootstocks to choose from.

Choose a variety that suits your taste preferences and then choose a rootstock suitable for your site.

Semi-dwarfing rootstocks such as MM106 and M26 produce trees that grow only about fifteen feet high and wide and so are suitable where space is limited. Where space is even more cramped, or for growing in pots, plant a tree grafted onto fully dwarfing rootstock such as M27 or P22. Both rootstocks produce trees maturing at only six feet high. (Some varieties— 'Garden Delicious,' for example—are genetic dwarfs, naturally small without being grafted on a dwarfing rootstock.) Dwarfing or semi-dwarfing rootstocks are useful even if you have abundant space but you would like to have more trees, each a different variety.

*Dwarf apple orchard on M27 rootstocks*

The rootstocks mentioned are only some of those available. Rootstocks also have been selected for their ability to tolerate adverse soil conditions. MM111, for instance, is adapted to growing in dry soils.

Below is a sampling of apple varieties, grouped by characteristics. Keep in mind that just because a variety is listed in one category does not mean that it does not also have other qualities. 'Esopus Spitzenberg,' for example, is an excellent apple to bite into when fresh but is also a good cooking apple. 'Liberty' is very disease-resistant but also tastes good.

**Apples for warm regions ("low chill" varieties):**
'Beverly Hills': The fruits are small to medium size, with a good-quality, tart flesh that resembles 'McIntosh.' The skin is pale yellow splashed with red. This variety dislikes hot weather and is moderately well suited to the coastal climate of southern California.

'Ein Sheimer': This fruit is similar to 'Golden Delicious,' with a yellow skin having a slight blush and a crisp, sweet-tart flesh. This variety is resistant to scab and is well suited to hot climates.

'Winter Banana': This beautiful fruit—clear yellow with a pinkish blush—was once widely used in fruit baskets. The flavor is aromatic and the slightly yellowish flesh does indeed have hint of banana flavor. This variety stores well.

**Best-tasting for fresh eating:**
'Ashmead's Kernel': Two hundred years ago Dr. Ashmead of Gloucester, England, found this apple growing in his garden. The fruit has a russeted golden brown skin, with a reddish bronze cheek. Inside, the flesh is aromatic, sweet, and crisp. This variety ripens late and keeps well.

ASHMEAD'S KERNEL

'Cornish Gilliflower': This fruit was discovered in a cottage garden in Cornwall, England, about two hundred years ago. It is an ugly apple, russeted brownish red over an olive base. But its flavor is supreme—a rich sweetness with a hint of clove.

'Cox's Orange Pippin': This sprightly and distinctively flavored fruit has a skin that is orange and red washed with carmine over a yellow background. Yields can be low, and the fruit tends to be small, sometimes sunburning or cracking. In spite of these problems, this variety remains the most popular apple in England.

ESOPUS SPITZENBERG          COX'S ORANGE PIPPIN

'Esopus Spitzenberg': This old American variety ripens in fall to a rich, somewhat tart flavor. The crisp fruit has a yellow skin covered with a mix of bright and dark red. In cold storage the fruit keeps well until spring. 'Spitz' was the favorite apple of Thomas Jefferson.

*'Esopus Spitzenberg' apple*

GALA

'Gala': This juicy, sweet apple is considered one of the best early apples. The skin is beautiful golden yellow with a pink-orange blush. In contrast to most other early apples, 'Gala' stores well, until Christmas. It bears at a very early age.

'Jonagold': This variety is a hybrid of the sprightly 'Jonathan' and the aromatic 'Golden Delicious.' The fruits are crisp and explode in your mouth with juice and flavor when you bite into them. The skin is yellow with a splash of light scarlet. 'Jonagold' has infertile pollen and so cannot pollinate other apple varieties.

'Mutsu': This large, round, yellow apple has a delicate, spicy flavor. The texture is pleasantly coarse, reminiscent of biting into a snowball.

**Apples for cooking:**
'Cortland': This variety has a dark red skin and a snow-white flesh that does not discolor on exposure to air. The fruit is excellent for eating fresh and makes an especially fine applesauce. 'Cortland' does not store well.

'Rhode Island Greening': This old American apple was once widely grown in the Northeast. It is a reliable and productive bearer. The grass-green fruit turns a little yellow as it ripens. Some people enjoy the fully ripe fruits fresh, and they are also good dried.

**Disease-resistant apple varieties:**
'Chehalis': This variety is similar to 'Golden Delicious' but is crisper and ripens in summer. It is not resistant to rust disease.

'Dayton': This red apple has a spicy, mildly tart flavor and stores for six weeks. It is not resistant to rust disease.

'Liberty': This large red apple has a delicious sweet-tart flavor, similar to 'Macoun.' The flavor improves in storage. This variety bears at an early age and is productive.

*'Liberty' apples on tree*

'Redfree': This summer apple has medium-sized, glossy red fruits with crisp, juicy flesh. It ripens unevenly and can be stored for up to two months.

**Apples for very cold climates:**
'Centennial': This variety is a sweet crabapple having bright red fruits. The fruit tends to be borne in alternate years.

'Haralson': This fruit has a crisp, juicy, sweet-tart flesh within a red and yellow skin. 'Haralson' bears at an early age and stores well through the winter.

HARALSON

'Wealthy': For best flavor, pick the tart, very juicy fruits when they are fully ripe, then eat them almost immediately. The tree is precocious and productive and, with its long bloom period, a good pollinator.

# APRICOT
*(Prunus Armeniaca var. Armeniaca)*

**Pollination needs:** Most, but not all, varieties are self-fruitful.

**Sun/shade requirements:** Full sun

**Zones:** 5 to 9

*Apricot fruit*

Apricot is difficult to grow over much of the country. The very early blossoms often succumb to late spring frosts; insects and diseases take their toll; and fluctuating winter temperatures weaken or kill trees. Apricots can be grown in Zones 5 to 9 but within this climate range do best where winters are steadily cold, springs are steadily warm, and the humidity during the growing season is low. Even if you live where conditions are not ideal for apricot, the tree-ripe fruit is so rich and sweet that you may want to give it a try if you resign yourself to inconsistent crops. Even if you lose a crop, the early, pinkish white blossoms are welcome and look stunning against the stark, late-winter landscape.

## Growing

For success with apricots, site the tree with care. The plants require full sunlight and a microclimate absent from frosts once growth begins. Apricots require perfect drainage but are otherwise not finicky about soil. The plants grow well over a wide pH range. With good care, the shoots on an apricot tree should grow twelve to eighteen inches each year.

Prune your apricot annually, beginning the year you set your tree in the ground, preferably just as growth begins in spring. Train your young tree to either an open-center or modified-central-leader form. Apricot trees bear fruit on short-lived spurs on older branches and directly on younger branches and therefore need pruning to stimulate growth of a moderate amount of renewal wood each season. Use heading and thinning cuts to remove old wood that is no longer fruitful and to stimulate new growth for the following season's fruit. Keep the branches open so that they are all bathed in light.

If you are fortunate enough to have a very heavy set of fruit on your tree, thin them to about two inches apart. Otherwise, no thinning is necessary.

## Pest Damage and Identification

Plum curculio attacks apricot in the eastern United States, leaving a crescent-shaped scar early in the season. Damaged fruit that remains on the tree is subject to disease, but most of these fruits will fall to the ground.

The other important insect pest of apricot, the oriental fruit moth, makes a large hole as it enters a fruit or else causes wilting of stem tips when it bores into a stem. Besides synthetic insecticides, this pest also has been controlled with repeated sprays of Bt, with *Macrocentrus* parasite, and in commercial orchards, with pheromones.

Borers also attack apricot, causing dieback of individual branches or whole trees. Trees whose bark has not been damaged by winter injury or lawnmowers and which are growing vigorously are less apt to be attacked.

On fruits the disease brown rot begins as a circular brown area that eventually envelops the whole fruit, which turns fuzzy gray. Besides spraying, picking off dried-up "mummies," on which the disease over-winters, and planting resistant varieties such as 'Alfred,' 'Harcot,' and 'Harlayne' help control this disease.

Various bark cankers (sunken, dark lesions) also can weaken or kill an apricot tree. A tree well hardened for winter, with a trunk that is protected from winter sun with a coating of white latex paint, is less apt to develop bark cankers.

## Selection of Plants

Many varieties of apricot exist, with varying adaptations, so choose varieties suited to your region.

'Alfred': Introduced in New York, this variety is adapted to eastern growing conditions. The tree is productive, cold-hardy, and vigorous. The medium-sized fruit is resistant to brown rot and has a sweet, rich flavor. Sometimes the skin will have a pink blush.

'Blenheim': This old English variety is grown commercially in California for drying and canning. The tree requires only moderate chilling and therefore blooms early, producing pale orange fruits that are sweet, aromatic, and very juicy.

'Harglow': Introduced by the Harrow Research Station in Canada, this tree is compact and blooms relatively late. The pure orange fruits are resistant to cracking and disease. They taste good fresh or cooked.

'Hargrand': Introduced by the Harrow Research Station in Canada, this variety is productive and cold-hardy. The fruits are very large and orange, with a speckled blush and good flavor. They resist cracking and are somewhat resistant to disease.

'Harlayne': This Canadian variety is cold-hardy and produces medium-sized fruits with a red blush. The fruit has a good flavor and texture and is moderately resistant to diseases.

'Manchu': This very cold-hardy and productive variety of apricot was developed in South Dakota in the 1930s. The fruit is large and yellow and best used for cooking.

'Moorpark': This old English variety is now grown commercially in California for fresh eating, canning, and drying. The blossoms are early. The fruits are very large, juicy, and sweet and ripen over a long period of time.

'Perfection': The tree is hardy but has a low chilling requirement and therefore blossoms early. This variety also needs cross-pollination. The fruit is large but only mediocre in quality.

'Scout': This Canadian variety is both cold-hardy and vigorous. The smooth, mild, sweet fruits are good for eating fresh as well as for making into jam.

# BLACKBERRY

*(Rubus spp.)*

**Pollination needs:** Self-fruitful.

**Sun/shade requirements:** Full sun to partial shade

**Zones:** 5 to 8

In the wild, blackberries are found growing almost everywhere in North America. Some of these wild plants creep along the ground, while others grow upright like small trees. These growth habits have been bred into cultivated blackberries, so you can choose from erect as well as semi-trailing and trailing varieties. Scientists have even bred out the ominous thorns from plants to create a few thornless varieties.

Blackberries have perennial roots but biennial canes. Each cane bears fruits in its second season, then dies. In any growing season, as older canes are fruiting, then dying, new ones are making their first season of growth. Your planting, therefore, has both one- and two-year-old canes, giving you an annual harvest.

You should reap your first harvest the year after planting. Because blackberries usually pick up virus diseases from wild plants, plan on replanting at a new location every decade or so—with certified virus-free plants, of course.

## Growing

Although wild blackberries often are found growing along the partially shaded edges of woods, the plants fruit best in full sunlight. Try to select a location as far as possible from wild brambles to avoid the spread of disease to your planting. The soil should be well drained and rich in organic matter.

Spacing of the blackberries depends on the type of blackberry that you plant and your training system. Erect blackberries are self-supporting, but other types are easier to manage if you train them on a trellis. Erect a trellis by sinking two sturdy posts deep into the ground at the ends of your row, then stringing two wires between them—one three feet above ground and the other five feet above ground.

You also can grow blackberries in "hills," sinking posts in the ground six feet apart each way, then setting one plant at each post. As the canes grow, tie them up to the post.

Prune blackberries twice each year. On all but the trailing types, do your first pruning in summer, pinching out the tips of new canes just as they reach a height of three feet. Pinching causes the canes to branch. Not all canes reach this height simultaneously, so look over your plants every couple of weeks to see if any canes need pinching.

Prune again in late winter, just before growth begins. First cut away at ground level any cane that bore fruit the previous season—these canes are dead anyway. (You also can prune away these canes in summer, right after they finish fruiting.) Next, thin out new canes, which will fruit next season, saving those that are thickest but leaving no more than six per clump. Finally, shorten lateral branches on the canes that you save to about eighteen inches, shorter if a cane is weakly and longer if it is vigorous.

## Pest Damage and Identification

Dying canes (except those that naturally die after they finish fruiting) could be the result of borers, verticillium wilt, or crown gall. Borers are least likely to attack vigorous canes, but in any case, diligently cut away and destroy canes that have been attacked. Verticillium is a soil-borne disease that is best avoided by not planting blackberries where tomatoes, peppers, eggplants, raspberries, or other hosts for the disease have recently grown. Varieties such as 'Logan,' 'Marion,' and 'Olallie' are resistant to verticillium. Avoid crown gall by not injuring the crowns of plants and sterilizing pruning or digging tools when working with the plants.

Bright orange pustules on the leaves in spring are signs of orange rust disease. Immediately dig up and destroy infected plants. Ignore plants having orange pustules late in the season—this is a different disease and does little damage.

Purple spots on leaves or canes indicate anthracnose or leaf and cane spot disease. Plants in a well-pruned patch dry off quickly, especially if the site is in full sunlight, and so are less likely to become diseased. Other ways to thwart these diseases include using sulfur sprays and planting resistant varieties such as 'Black Satin,' 'Dirksen Thornless,' and 'Flint.'

Reddish, twisted flowers that do not develop into fruits are symptoms of double blossom disease, common in the Southeast. Cut all infected canes to the ground right after harvest and destroy them. 'Florgrand,' 'Himalaya,' and 'Humble' are resistant to this disease.

Fruits that remain red, hard, and sour have been attacked by redberry mites.

## Selection of Plants

Blackberries vary in their adaptation to climate, so seek information from nurseries and the local Cooperative Extension for varieties best suited to your region. Once you have some choices, narrow your selection on the basis of growth habit, thorniness, and, of course, flavor.

'Black Satin': The plant is semi-trailing, thornless, and cold-hardy. The large fruits turn dull black when ripe.

'Boysenberry': The fruits are very large and dark maroon, soft and very juicy, and have a tangy flavor. They do not store well. The vine is trailing and adapted to Pacific Coast states but is not hardy in cold areas. A few different clones exist, varying in vigor, cold-hardiness, and thorniness. This variety may be the same as 'Burbank's Phenomenal.'

'Cherokee': This variety is erect, moderately thorny, and productive. The fruits are large and firm and taste good fresh, frozen, and canned. The plant is hardy in Zones 5 to 8.

'Chester': One of the hardiest (to about −10° F) and most productive of the thornless varieties, 'Chester' has large sweet fruits.

*'Chester' blackberry fruits*

'Cheyenne': Plants are erect and moderately thorny, holding their fruits high above the ground. Although susceptible to rosette disease, plants are resistant to orange rust and moderately resistant to anthracnose. The fruits are very large, with good flavor.

'Darrow': This is one of the most cold-hardy blackberries (to about −22° F). The upright plants are very thorny. The fruits are large, with excellent flavor.

'Hall': This variety is a sibling of 'Black Satin' but is more cold-hardy and has sweeter and firmer fruits. The fruits are also large. The plants are semi-trailing, thornless, and productive. This variety is adapted to both the East and the Northwest.

'Loganberry': This hybrid of a California blackberry and 'Antwerp' red raspberry is thornless and moderately cold-hardy. The berries are light red and tart when ripe, excellent for jam. Early in this century this was the most popular blackberry in California.

'Olallie': The hybrid of a wild blackberry, 'Youngberry,' and 'Loganberry,' this variety bears fruits that are large, firm, and shiny black. The canes are trailing. 'Olallie' is adapted to growing in California, western Oregon, and along the Gulf Coast. The plant has a low chilling requirement.

'Tayberry': This hybrid of blackberry and raspberry produces large fruits that retain their core, like blackberries. The plants are semi-erect, productive, and hardy to about −15° F.

# BLUEBERRY
*(Vaccinium spp.)*

**Pollination needs:** Cross-pollination not required, but increases yield and size of fruit.

**Sun/shade requirements:** Full sun to partial shade

**Zones:** Highbush varieties, Zones 4 to 7; rabbiteye varieties, Zones 7 to 9; lowbush varieties, Zones 3 to 7

Although blueberries have been harvested from the wild for centuries, this native fruit has been cultivated only since the beginning of the twentieth century. Blueberries not only provide fruit but are also very ornamental plants for the landscape, especially in the fall, when their leaves turn red.

Three species of blueberry are grown for their fruits. The most common, the one you see fresh on grocers' shelves, is the highbush blueberry (*V. corymbosum*), native to the East Coast, with bushes that grow about six feet high. The Southeast is the home of the rabbiteye blueberry (*V. asheii*), which grows fifteen or more feet high. These berries are not quite as tasty as those of highbush blueberry, but the plants are better able to tolerate hot weather and drier soils.

The third species, the lowbush blueberry (*V. angustifolium*), grows wild on rocky hillsides in the Northeast. The plants spread underground to form a solid mat with stems rising less than eighteen inches high. The berries are sweet and often pale blue, with a powdery "bloom." Lowbush blueberries are the cold-hardiest of the three species.

Highbush and lowbush blueberries have been hybridized to produce medium-sized plants, called half-highs, that are cold-hardy and produce tasty fruits. Half-highs generally do not spread underground as do lowbush blueberries.

## Growing

Although wild blueberries commonly grow in the dappled shade of forests, the plants fruit best in full sunlight.

The key to success with blueberries is to make the soil to their liking. The soil must be very acidic, well drained, and high in organic matter. Test the pH before you plant, then add sulfur and/or acidic peat moss to bring the pH to between 4 and 5. The peat moss is also useful for adding organic matter to the soil; even if the pH is already sufficiently acidic, mix a bucketful of peat moss with the soil where you plant.

If your soil is very alkaline, make soil for blueberries rather than try to acidify existing soil. Dig out a hole two feet deep and six feet wide, and replace the soil with a mixture of equal parts peat moss and either sand or perlite.

After you plant, mulch the ground around the bushes with a few inches of leaves, sawdust, hay, or other organic material. As the mulch decomposes, the bottom layer will enrich the soil with humus, as well as fend off weeds and protect the very shallow roots from the heat of summer sun.

Over the years, maintain soil acidity with periodic sprinklings of sulfur and by using nitrogen fertilizers such as ammonium sulfate, soybean meal, or cottonseed meal. Yellowing of the youngest leaves, with the veins remaining green, usually indicates that the soil is not acidic enough.

Highbush blueberries need no pruning until they are four years old. Then begin pruning by cutting away the oldest stems near the ground to make room for new stems. Next, shorten stems that droop to the ground, twiggy growth, and any stems overcrowded in the center of the bush. Finally, shorten stems having many plump fruit buds at their ends so that only about five fruit buds remain.

*Cutting off excess fruit buds from blueberry stem*

Prune rabbiteye blueberries the same way as described for highbush blueberries, but less severely.

Prune lowbush blueberries by clipping or mowing the stems to the ground in the winter every second or third year. The plants do not bear fruits the season following mowing, so to maintain a continuous harvest from a bed, mow a different half of the bed every second year or a different third of the bed every third year.

## Pest Damage and Identification

The major pest problem in growing blueberries is birds, and the only reliable protection against birds is netting. The ideal is a walk-in cage, either temporary or permanent, erected over the whole planting.

Usually you can grow blueberries with no other pest problems. Occasionally, though, blueberry maggots—small white "worms"—may turn up in the fruits. Thorough harvest of all sound and infested fruits helps thwart this pest, as does a thick mulch, renewed yearly. You also can trap this pest by hanging red spheres coated with sticky Tangletrap among the bushes just before the first berries ripen. Hang one trap for every thirty square feet of planting.

Mummy berry is a disease that causes fruits to fall before they ripen or to turn reddish, then tan, and shrivel into hard mummies on the plant. Picking off all fruits, good and bad, helps control this disease, as does an annual renewal of mulch, which buries infected fruits you miss. 'Bluetta,' 'Burlington,' 'Darrow,' 'Dixi,' 'Jersey,' and 'Rubel' are resistant to mummy berry.

Brown or gray mold on the fruits or leaves is a sign of botrytis blight, a disease most prevalent in cool, wet weather. Avoid botrytis blight by disposing of dead twigs and rotted fruits, pruning to keep a bush open to the drying effects of air and sun, and being careful not to overstimulate growth with fertilizer.

## Selection of Plants

### Highbush varieties:

'Atlantic': The medium blue fruits are large and good quality, ripening late in the season. The plant is vigorous and productive and has a spreading growth habit.

*Netted blueberry planting*

'Berkeley': The light blue fruits are very large, with good flavor. They ripen midseason. Plants are productive.

*'Blueray' blueberry fruits*

'Bluechip': This excellent variety has good flavor, size, and color. The bush is productive and upright.

'Bluecrop': The light blue berries are large, firm, and good quality. The fruit ripens midseason, sometimes dropping when ripe. The bush is vigorous, productive, and more drought-resistant than other varieties. A cold-hardy variety.

'Earliblue': The fruits are large and ripen early. The plant is vigorous and somewhat upright, with good cold-hardiness.

'Jersey': The fruit ripens late and is large, with fair flavor. The bush is cold-hardy and upright growing.

'Wolcott': The fruit is dark blue and good quality, ripening early and over a short period of time. The berries hang in loose clusters. The plants are upright and productive.

**Rabbiteye varieties:**
'Aliceblue': The fruit is small and flavorful. The plant is tall and spreading. This variety needs cross-pollination to set fruit.

'Beckyblue': The fruits are medium-sized, with good flavor. The plant is tall and spreading. This variety needs cross-pollination.

**Lowbush varieties:**
'Augusta': Fruits of this variety are small and powdery blue, with fair flavor. The plant grows only six inches high. This variety needs cross-pollination for best crops.

'Brunswick': The fruits are about a half-inch across, medium blue in color, with excellent flavor. Production is variable on these foot-high bushes.

'Chignecto': These moderately productive, vigorous plants produce berries with excellent flavor.

**Half-high varieties:**

'Northblue': The fruits are large and dark blue, slightly tart but with good flavor. Plants are productive, growing twenty to thirty inches tall.

'Northsky': The fruits are small and sky blue in color, with good flavor. Plants grow ten to twenty inches tall.

'Patriot': A cold-hardy variety with a growth habit close to that of highbush blueberry. The fruits are large and have very good flavor.

'Tophat': The fruits are medium to large in size, with only mediocre flavor. The bush is ornamental, though, growing about two feet high and wide, with dainty leaves that turn red in fall.

# CHERRY

*(Prunus* spp.*)*

**Pollination needs:** Most varieties of sweet cherry and Nanking cherry require cross-pollination; tart cherries are self-fruitful.

**Sun/shade requirements:** Full sun

**Zones:** Tart cherry, Zones 4 to 8; sweet cherry, Zones 5 to 9

Take your pick: do you want sweet cherries (*P. avium*) to pluck right from the tree and eat fresh, or do you want tart cherries (*P. cerasus*) for piping-hot cherry pie? Taste is not the only difference between these two cherries. Sweet cherries are generally borne on large upright trees, while tart cherries are borne on small spreading trees. And tart cherries are the easier of the two types to grow, being more tolerant of cold in winter and less finicky about soil.

*Tart cherry fruits—'Montmorency'*

*Sweet cherry fruits—'Bing'*

In the case of cherries, you can have your cake (pie?) and eat it too, by growing a third type of cherry—duke cherries (*P.* × *effusus*). Duke cherries, although less commonly available from nurseries, are natural hybrids of sweet and tart cherries, combining their qualities.

And there are even other cherries to choose from. The Nanking cherry (*P. tomentosa*), a bush native to the hills of Manchuria, can be used as an ornamental fruiting hedge and is very cold-hardy, tolerant of drought, and pest-resistant. The fruits are small and only slightly tart. The western sand cherry (*P. Besseyi*), sometimes known as Hansen's bush cherry, is another cold-hardy and somewhat drought-tolerant cherry. The fruits are good for jam but not for fresh eating.

## Growing

Especially with sweet cherries, choose your site with care. Good soil drainage and full sunlight are essential. Sweet cherries blossom very early in spring, so choose a site not prone to late frosts.

Sweet and tart cherry trees differ in their growth habits. Train a sweet cherry tree to either a central-leader or modified-central-leader shape. The spreading habit of a tart cherry tree makes it a natural for an open-center form.

Once trained and beginning to bear fruit, both sweet cherry and tart cherry trees need only light annual pruning. Both trees bear fruit on both young wood and on spurs on older wood, so prune to stimulate just a little annual growth. Also remove any

*Nanking cherry fruits*

diseased, damaged, or misplaced branches. Your combined pruning and fertilization should result in about eighteen inches of new shoot growth each season.

## Pest Damage and Identification

Birds are a major pest problem on cherries, especially sweet cherries. Netting a tree is the only sure way to keep birds at bay, but this is not feasible for a large tree. A less effective alternative is to drape a tree with black cotton thread, which bothers the birds. Do this by tossing a spool back and forth among the branches while holding onto the free end of the thread as it unrolls from the spool. Yellow varieties are less attractive to birds than are red varieties.

Crescent-shaped scars on young fruits, commonly causing them to drop, indicate the work of plum curculio, an insect pest prevalent east of the Rocky Mountains.

Fruits that become covered with a brown or gray fuzz have been attacked by brown rot disease. Plants are most susceptible to infection during bloom and just before fruits begin to ripen, especially if the fruits have been damaged (by plum curculio, for example). Cleaning up hard, dry mummies still hanging in the tree from the previous year's disease removes one source of infection, but spraying is also often necesssary.

If you find small white "worms" in your cherries, they are the larvae of the cherry fruit fly. Instead of spraying, you can trap these flies on sticky red traps hung in the trees in May.

Black cherry aphids cause cherry leaves to twist and curl. Masking tape wrapped around the trunk and coated with Tangletrap stops aphid-herding ants from climbing up and down the trunk.

The rainy spring weather that is common in the East and the Midwest commonly causes cherry leaf spot disease, symptoms of which are purple spots on the upper surfaces of the leaves. Because the disease overwinters in old leaves, raking them up in autumn removes sources of infection. Resistance to this disease is found in tart cherry varieties such as 'Northstar' and 'Meteor' and sweet cherry varieties such as 'Hedelfingen,' 'Lambert,' and 'Schmidt Bigarreau.'

Dry weather can bring on a different disease, powdery mildew, which covers the leaves with a powdery white coating. As the disease worsens, leaves also become twisted.

## Selection of Plants

In addition to considering whether you want to grow sweet and/or tart cherries, also consider the eventual size of the tree. Although sweet cherries do grow into large trees, dwarf rootstocks are becoming available, and some varieties, such as 'Compact Stella,' are naturally dwarf.

Most sweet cherry varieties need cross-pollination, but not all varieties are compatible, so check nursery catalogs to make sure the varieties that you want can pollinate each other.

'Bing': This is the standard sweet cherry, with large dark red, heart-shaped fruits having a firm, meaty flesh. The fruit cracks in wet weather. The tree is only moderately hardy and a shy bearer.

'Black Tartarian': The fruits are medium to large, dark red, and heart shaped. The flesh is tender, very juicy, and sweet. The tree is productive.

'Compact Stella': This sweet cherry has medium-large fruits that are dark red, firm, but of only fair flavor. The tree is semi-dwarf and moderately hardy. This variety is reliably productive and self-fertile.

'Emperor Francis': This variety of sweet cherry has delicious yellow fruits that resist cracking. The tree is productive and cold-hardy.

'Kristin': The large sweet fruit has a tender, dark red skin and firm, red flesh. The fruit tends to crack if rainy weather occurs near harvest. The tree is cold-hardy.

'May Duke': This duke cherry has an excellent tart flavor. The dark red fruits ripen early and are excellent for cooking also. Cross-pollinate this variety with a sweet or sour cherry.

'Montmorency': This is the standard pie cherry, with bright red fruits having rich, tangy flavor. The fruits are firm and resist cracking.

'Northstar': These tart cherries are borne on a naturally small tree. They are light red, resist cracking, and hang for a long time after they are ripe. The plant is very cold-hardy and somewhat resistant to cherry leaf spot and brown rot.

'Sodus': The large sweet fruits have light red skin and a firm, white flesh. The fruits resist cracking.

'Sweet Ann': This yellow sweet cherry has a red blush, a firm texture, and resists cracking. The flavor is excellent. The tree is cold-hardy and somewhat resistant to spring frosts.

'Van': This sweet cherry has dark red skin and firm texture, very similar in flavor to 'Bing.' The fruit is somewhat resistant to cracking.

# CURRANTS
*(Ribes spp.)*

**Pollination needs:**  Self-fertile, except for certain varieties of black currant

**Sun/shade requirements:**  Partial shade

**Zones:**  3 to 6

Three types of currants are commonly grown for their fruits: the red currant, the white currant, and the European black currant. Red and white currants are botanically the same, differing primarily in color. All three currants make tasty jellies, jams, and beverages. What's more, currants are among the easiest of fruits to grow. A currant enthusiast of the last century wrote that "the currant takes its place among fruits that the mule occupies among draft animals—being modest in its demands as to feed, shelter, and care, yet doing good service."

*'Consort' black currant fruits*

Currants are fruits of northern climates. The bush, growing about four feet high and wide, tolerates extremely cold winters and enjoys cool summers. The fruits ripen early in the season, just after strawberries.

## Growing

Currant bushes enjoy the coolness of a north slope or partial shade, especially where summers get very hot. No matter where you plant them, but especially in full sunlight, keep the ground beneath the bushes blan-keted with a thick organic mulch such as straw or leaves. The plants appreciate the cool soil beneath the mulch, which also suppresses weeds and conserves soil moisture.

Grow currants either as individual bushes or as a continuous hedge. Space individual bushes six feet apart. For a hedge, set plants three feet apart. Right after you plant a bush, cut back all but three stems to ground level to stimulate a supply of new wood.

You will get best production from a bush if you prune it every year. Cut away old wood at ground level and thin out new growth so that it is not crowded, to about six stems. Red and white currants fruit mostly on one-, two-, and three-year-old stems, so always remove stems more than three years old. Black currants fruit mostly on one-year-old stems, so cut back stems more than one year old either to the ground or to a vigorous branch.

*Red currant before pruning*

*Red currant after pruning*

## Pest Damage and Identification

Currants usually can be grown without any need for pest control. In some regions, powdery mildew disease may cover new leaves with a white coating. Or some of the leaves on a red or white currant bush may turn reddish and puckery, indicating aphid attack. Aphids usually depart or die before causing excessive damage. Borers may cause an occasional branch to die back. Cut back any infested stem below where the borer entered (indicated by a swelling on the stem) and destroy the stem.

White pine blister rust is a disease requiring two plants to complete its life cycle: a susceptible variety of currant (or gooseberry) and a white pine (or other five-needled pine). The disease causes only minor damage to currants but can kill pines. Red and white currants are not very susceptible to white pine blister rust. Black currants are very susceptible, except for resistant varieties such as 'Consort' and 'Coronet.'

## Selection of Plants

'Ben Sarek': This black currant yields tasty berries that are borne prolifically on a dwarf bush. The plants resist mildew.

'Boskoop': This old Dutch variety produces large good-flavored black currants.

'Consort': This black currant is immune to white pine blister rust.

'Jonkheer van Tets': Because of its productivity and ease of picking, this red currant is a leading commercial red currant in Holland. The fruit is large, with fair flavor.

'Red Lake': One of the best varieties of red currant, this variety is productive and bears fruits with good size and excellent flavor.

*'Red Lake' red currant fruits*

*'White Imperial' white currant fruits*

'White Imperial': This white currant has excellent flavor. The bush is upright growing and small in size.

'Wilder': This old American red currant bears large flavorful berries. Prune this variety more severely than other varieties.

# ELDERBERRY
*(Sambucus canadensis)*

**Pollination needs:**  Partially self-fruitful

**Sun/shade requirements:**  Full sun to partial shade

**Zones:**  2 to 9

Elderberries grow wild throughout eastern portions of the United States. The bushes are especially prominent, as their flat-topped clusters of creamy white flowers dot the borders of woodlands well after the floral show from other spring-flowering shrubs has long faded. Elderberry flowers, incidentally, are edible and sometimes used in making wine.

Elderberries ripen in summer, bearing heavy clusters of purplish black fruits that weight the branches downward. The fruits are insipid when eaten fresh but make a tasty jam, jelly, or pie.

## Growing
Elderberries tolerate a wide range of soils, as well as sites in full sun or partial shade. The bushes grow about eight feet tall, sending out many suckers from the ground. Space plants eight feet apart unless you want to grow elderberries as a hedge.

For maximum productivity, prune elderberry bushes annually, in winter. Do this by cutting away old wood at or near ground level in order to make room for new growth. Also thin out some new shoots, saving those that are most vigorous.

## Pest Damage and Identification

Birds may eat some of the berries. Otherwise, elderberry is free from pests.

## Selection of Plants

'Adams': This plant bears large clusters of large fruits.

'Scotia': This variety has especially good flavor.

'York': This variety yields the largest elderberries of all, on a bush that is both large and productive. The berries ripen late.

# FIG

*(Ficus carica)*

**Pollination needs:** Not required for most varieties (the commercial variety 'Calimyrna' has special pollination needs)

**Sun/shade requirements:** Full sun

**Zones:** 8 to 10

Figs are one of the oldest cultivated fruits, originating in the Mideast. Despite the plant's subtropical origins, you can grow a fig—one way or another—no matter where you live. This is because some varieties bear fruit not only on old wood but also on new shoots. So if your winters kill the tree to the ground, you may still get a crop on new growth. Figs are easy to raise in pots, which is how you can grow them even if your winters are bitterly cold and your summers are not long enough to ripen fruits that form on new shoots.

## Growing

During the growing season, figs thrive in heat and sunlight. In northern areas, grow your figs near a south-facing wall in order to capture heat and, if the wall is a light color, to reflect additional light to the plants.

Northern gardeners have devised a number of ways to eke fruit from a fig tree. If your winters are cold enough to kill stems to the ground, yet your growing season is not long enough to ripen figs on new shoots, try protecting the branches in winter. One way is to tie the stems together, then surround them with a wire cage, into which you stuff dry leaves. Top the cage with a waterproof covering. For even more protection, slowly bend the plant to the ground in autumn, after all its leaves have fallen. Cutting the roots on the side opposite the direction you bend the plant makes bending easier and causes little harm to the plant. Weight the plant to the ground with some rocks or cinder blocks, then cover the plant with some insulating material such as dry leaves or straw, some plastic to keep water off, then additional leaves or straw. The depth of covering needed depends on how cold temperatures become in winter.

Protect a potted fig tree from winter cold by moving the whole plant to shelter. As long as a plant is leafless, it does not need light. Keep the plant dormant and leafless through winter by storing it in a cool spot, ideally between 30° and 50° F. Repot your fig every year, cutting back its roots, then putting it back into the same pot after the plant reaches full size.

Outdoors in the ground or indoors in a pot, train a fig either as a bush with many stems or as a tree with a single trunk. The bush form is more practical if winter cold will occasionally kill part or all of your plant back to ground level. Figs fruit fairly well with little or no pruning, but pruning can increase production. Certain varieties, such as 'Adriatic,' 'Alma,' 'Brown Turkey,' 'Celeste,' 'Everbearing,' 'Kadota,' and 'Magnolia,' fruit mostly on new shoots, so prune these varieties severely enough to stimulate ample new growth each year. 'King,' 'Mission,' 'Tena,' and 'Ventura' are examples of varieties that fruit on one-year-old stems and new shoots and therefore should not be pruned severely.

## Pest Damage and Identification

Figs have few pest problems. Where soil nematodes are a problem, as in certain areas of the Southeast, plant figs near buildings. The roots grow underneath the building, out of reach of the pests. Adding quantities of organic matter to the soil also helps against nematodes.

Where dried fruit beetles are present, grow varieties such as 'Eastern Brown Turkey' and 'Celeste,' which have small "eyes" (the opening at the far end of the fruit) through which the beetles cannot enter.

## Selection of Plants

'Adriatic': This old variety, probably originating in Italy, bears two crops per season, one on old wood and one on new wood. The fruit has greenish yellow skin and strawberry-red pulp. The plant is adapted for growing in cool coastal areas.

'Brown Turkey': Two different varieties parade under this name, one in the East and one in the West. Both varieties look similar, have delicious flavor, and respond well to heavy pruning. Eastern 'Brown Turkey' is sometimes called 'Everbearing' or 'Texas Everbearing' and is a relatively cold-hardy variety.

*'Brown Turkey' fig on tree*

'Celeste': The fruits are small, sweet, and light brown to violet in color. Prune this variety severely. 'Celeste' is fairly cold-hardy and fruits well in the Southeast.

'Kadota': The fruits of this ancient variety are large, bright greenish yellow, and excellent for eating fresh or for canning. 'Kadota' has a large eye.

'King': The fruit has dark green skin and pink flesh, with excellent sweet flavor. It generally sets only on older wood, so do not prune severely. This variety is well suited to cool coastal climates.

# GOOSEBERRY

*(Ribes hirtellum* and *R. uva-crispa)*

**Pollination needs:**   Self-fertile

**Sun/shade requirements:**   Partial shade

**Zones:**   3 to 7

Gooseberries are a versatile fruit, delicious cooked into pies and jams and—yes!—even eaten fresh. So-called dessert varieties of gooseberry are sweet, comparable in flavor to grapes. In northern Europe, appreciation of the gooseberry spurred efforts of both amateur and professional breeders, so that now there are hundreds of gooseberry varieties.

The plants themselves are usually thorny bushes, some upright, some sprawling, depending on the variety. The fruits ripen soon after the last of June-bearing strawberries finish ripening.

*Bowl of many varieties of gooseberries*

## Growing

Gooseberries are plants of northern climates, tolerating cold winters and enjoying cool summers. If possible, especially in hot summer areas, plant the bushes on a cool northern slope or in the coolness of partial shade. After planting, blanket the ground with a mulch of leaves, straw, or other organic material, renewed as necessary to keep the soil cool, moist, and weed-free.

Gooseberries fruit mostly on stems that are one to three years old. Prune your bush each winter to leave about a half-dozen each of one-, two-, and three-year-old stems. Do this by cutting away at ground level any stems more than three years old and thinning out one-year-old stems so that only a half-dozen of the most upright ones remain. You can tell the age of a stem because it becomes darker and more peeling with time. Also shorten any stems that you leave whose tips droop to the ground.

## Pest Damage and Identification

A white powdery coating on young leaves is a sign of powdery mildew disease. Varieties such as 'Poorman,' 'Glendale,' 'Hinnonmakis Yellow,' and 'Lepaa Red' resist this disease.

Leaf spot diseases cause the leaves to develop dark spots, then to yellow and drop. Although diseased bushes become unsightly, they usually continue to bear reliable crops. If severe, spray.

Two insects occasionally appear: the gooseberry fruitworm, which bores into a fruit, causing it to color prematurely, and the imported currantworm, which strips the leaves just after they unfold. Both pests are easily controlled by sprays applied as soon as you notice damage.

## Selection of Plants

'Achilles': Fruits of this variety are among the tastiest and largest of dessert varieties of gooseberry. The ripe fruits are red or green, depending on the climate.

'Hinnonmakis Yellow': The yellow fruits have a delicious flavor that hints of apricot. The bushes are resistant to powdery mildew.

'Lepaa Red': This variety bears small red, somewhat tart fruits. The plant is very disease-resistant.

'Poorman': The fruits are medium-sized, pear shaped, and reddish, with a delicious sweet flavor. The plants are upright growing and only moderately thorny. This variety is resistant to mildew.

'Welcome': This sprawling bush bears medium-sized red fruits having a sweet-tart flavor. The plant is somewhat resistant to mildew.

*'Welcome' gooseberry fruits*

# GRAPES
*(Vitis* spp.)

**Pollination needs:** Except for certain varieties of muscadine grape, grapes are self-fertile.

**Sun/shade requirements:** Full sun

**Zones:** 3 to 10

Grapes are rather specific in their climatic requirements, but there are types and varieties adapted almost everywhere, from the cold reaches of the upper Midwest and New England to the sultry Southeast and the arid West. Vinifera grapes (*V. vinifera*, sometimes called European wine grapes) originated in eastern Europe and western Asia and are adapted to regions of our West having mild winters and hot, dry summers. 'Thompson Seedless' is typical of vinifera grapes, whose edible skins and sweet, crisp flesh make them ideal for fresh eating as well as for making into raisins and wine. American grapes (*V. labrusca*) are adapted to withstand cold winters, humid summers, and the indigenous pests found east of the Rockies, where these grapes are native. The "foxy" flavor (characteristic of "fox" grapes) and slipskin of American grapes is typified by 'Concord,' used fresh, for juice, and occasionally for wine. Grapes best adapted to the Southeast are those derived from native muscadine grapes (*V. rotundifolia*), which include varieties such as 'Scuppernong' and 'Hunt.' The large fruits of muscadine grapes are good fresh or made into wine.

Vinifera grapes and American grapes have been hybridized to combine the qualities of each species. The hybrids vary in cold-hardiness, pest tolerance, and, of course, fruit characteristics.

## Growing

A grapevine will produce tastier fruits and be healthier if you plant it in full sunlight. Grapes have far-reaching roots, so the ideal site also has deep, well-drained soil. Adjust planting distances according to the vigor of the varieties you plant.

For the best quality and ease of management, train your grapes when they are young and then prune them every year in late winter. Your grapes may bleed sap when you prune them, but this is harmless to the plants.

Grapes bear fruit only on canes, which are stems that grew the previous year. The goals of various methods of training and pruning grapes are the same: to leave a suitable number of canes, growing from the main trunk or permanent arms, for fruit the coming season, and to provide buds to grow into new, well-placed shoots to be saved for fruiting the following season. Three common methods for training grapes are head training, the four-arm Kniffin system, and the cordon system.

Use head training for grapes that fruit near the base of canes. Vinifera grapes generally bear fruit in this way (not 'Thompson Seedless,' however). Begin by training a single shoot, which will become the permanent trunk of the vine, up a stake about five feet high. Remove all other shoots. Cut off the tip of the shoot when it grows just above the top of the stake. Allow side buds along the upper one-third of the trunk to grow out into shoots to form a head. Each year prune canes growing from the head back to two or three buds. Always remove any shoots growing lower on the trunk.

Use the four-arm Kniffin system for most American grapes, some vinifera grapes, and muscadines. You will need to erect a trellis with two wires—at two feet and five feet above ground level—strung between two sturdy endposts. Train your young vine to a single trunk up to the top wire, then cut off the top and allow

*Grape arbor before pruning*

*Grape arbor after pruning*

four arms to grow, two in opposite directions at the level of each wire.

Annual pruning of the four-arm Kniffin system consists of four steps. First, mark four canes with a ribbon so that you do not accidentally cut them off. Choose two pencil-thick canes growing in either direction along the top and the bottom wires. Second, select wood that you will save as four renewal spurs, two near each wire. Renewal spur wood can be any age, as long as it has two or three plump buds near its base. Cut renewal spurs back to two buds, which will grow shoots, some of which you will save next year as fruiting canes. For the third step, cut off all growth except the four canes, the four renewal spurs, and the trunk. The final step in pruning is to shorten the fruiting canes that you saved to about ten buds each, more if last year's growth was vigorous, less if it was weak.

You can train and prune all three types of grapes to cordons, a system especially useful for plants grown on arbors. A cordon is a permanent arm growing off the trunk, or just a long, permanent extension of the trunk. Allow both canes and renewal spurs to grow off at intervals along the length of the cordon(s). If you are growing a variety that fruits mostly at the base of canes, prune to leave only short canes growing off each cordon.

In addition to keeping your vine within bounds and opening it up to light and air, pruning also removes potential fruits. Some varieties benefit from further fruit thinning. Do this by removing whole bunches and, especially where bunches are tight, cutting off some clusters of fruit within or at the end of a bunch.

## Pest Damage and Identification

Diseases threaten grapes, but you may be able to avoid problems by growing varieties adapted to your region. Four of the most common diseases are downy and powdery mildew, both of which cause white coatings on fruits; black rot, which starts as a pale brown circle on the fruit but eventually causes the whole fruit to shrivel and darken into a mummy; and botrytis bunch rot, which causes a fluffy, gray-brown coating on the fruit.

Insects sometimes cause their share of trouble. Some prominent ones include the grape berry moth, which occasionally makes holes in and feeds on fruits east of the Rocky Mountains; phylloxera, which can kill vinifera grapes as it feeds on their roots and causes telltale swellings; and Japanese beetles. Avoid phylloxera problems by planting resistant varieties, or susceptible varieties grafted onto resistant rootstocks.

## Selection of Plants

'Canadice': This vinifera-American hybrid bears seedless red fruits with the flavor of eastern grapes and an edible slipskin. The clusters are large and tight. The vines are winter-hardy. Do not let this variety bear too heavily, which it tends to do if allowed.

'Concord': This is the most familiar American grape, with blue-black fruits, a slipskin, and a foxy flavor. 'Concord' is adapted to growing over much of the country.

'Edelweiss': This American grape has a very sweet and foxy flavor. The berries are pale gold. The vine is cold-hardy and extremely vigorous.

*'Concord' grape*

*'Alden' grape*

*'Swenson' red grape*

'Flame Seedless': This vinifera grape has small seedless red fruits having crisp texture and excellent flavor. The fruit, borne in long, loose clusters, is good fresh and for raisins. The vines are moderately vigorous and productive.

'Hunt': This muscadine grape bears large black fruits that are very sweet. Provide a pollinator.

'Jumbo': This muscadine variety bears large purple fruits that ripen over a period of several weeks. The vine is very vigorous and productive, and adapted to growing throughout the Southeast. Provide a pollinator.

'Lakemont': This vinifera-American hybrid has medium-small sweet, seedless fruits with yellowish green skins. The vine is productive and moderately hardy (more so than its sibling 'Interlaken Seedless,' but less so than its sibling 'Himrod').

'Perlette': This seedless vinifera grape has a crisp flesh, a mild flavor, and a green skin. The fruits are small and ripen early, about a month before 'Thompson Seedless.' Prune to short canes.

'Reliance': This vinifera-American hybrid bears a pink fruit with a tender skin and a mild foxy flavor. The fruit stores well and the vine is cold-hardy, tolerant to disease, and adapted throughout the East.

'Scuppernong': This is the oldest muscadine variety. The large fruits have a thick, reddish bronze skin. The vine is vigorous and, with a pollinator, productive and reliable.

'Suffolk Red': This vinifera-American hybrid bears long, loose clusters of crisp, spicy-sweet, seedless red berries. The vine is only moderately hardy.

'Thompson Seedless': This vinifera grape has a mild, sweet flavor. Leave long canes when you prune. 'Thompson Seedless' does best in hot regions of the West.

'Vanessa Seedless': This seedless vinifera-American hybrid red grape has a crisp flesh and a mild, sweet flavor. The vine is moderately cold-hardy.

# JUJUBE
*(Ziziphus jujuba)*

**Pollination needs:** Needs are ill-defined; cross-pollination may or may not be needed, depending on the variety and the climate.

**Sun/shade requirements:** Full sun

**Zones:** 5 to 10

When just ripe, a jujube has the texture and flavor of a sweet apple. Left hanging on the tree a bit longer, the fruit lives up to one of its other names, Chinese date, as it shrivels, turns brown inside and out, and becomes concentrated in sugars. The fruits ripen in late summer or early autumn.

Because of its handsome, glossy green leaves, its fruit, and its adaptability to most soils, this medium-sized tree long ago traveled from its native home in China to southern Europe and then, in the nineteenth century, on to America. The tree has been very popular in China, so there are many varieties, some with such descriptive names as 'Dragon's Claw Jujube,' 'Tooth Jujube,' 'Bottle Jujube,' and 'Mellow Jujube.'

The plant tolerates cold to Zone 5 but fruits best where summers are very hot.

*Fresh and drying jujube fruits*

## Pest Damage and Identification

Jujubes have no significant pest problems.

## Selection of Plants

'Lang': The fruit of this variety is pear shaped, fairly large, and has a sweet flavor hinting of caramel.

'Li': These fruits are among the largest of any jujube varieties, round and two inches in diameter, with excellent flavor.

# JUNEBERRY

*(Amelanchier spp.)*

**Pollination needs:**  Self-fruitful

**Sun/shade requirements:**  Full sun to partial shade

**Zones:**  3 to 8

## Growing

Although needing a sufficiently long season and abundant sunlight to ripen its fruit, jujube will grow in almost any soil, whether wet or dry, hard packed or loose. The trees send up suckers, often a few feet from the mother plant, so plant the tree in lawn, where your mower can keep suckers in check. Control weeds and conserve soil water by mulching rather than by tilling right around your tree, because tilling encourages suckering.

The trees require no pruning or thinning of fruits.

The many species of native American juneberry, also known as serviceberry or shadbush, range in size from low shrubs to medium-sized trees. The plants are commonly grown as ornamentals, mostly for their early spring show of white or pink blossoms and for their blazing fall color. Juneberry fruits are usually the size and color of blueberries, and are juicy, with a sweet, almondy flavor. One bushy species, the saskatoon, is native to northwestern prairies and even grown commercially for its fruit.

*Juneberry fruits*

## Growing

Juneberry is an easy plant to grow, tolerating a wide range of soils and sunny or partially shaded sites. Tree species need little or no annual pruning, but suckering bushy species, such as the saskatoon, fruit best if pruned yearly, in winter. The best fruit is produced on wood one to four years old, so each year cut stems more than four years old to the ground. Also thin out the youngest stems if they crowd.

## Pest Damage and Identification

The major pests of juneberries are birds. Mature plants commonly yield enough berries for both human and feathered berry lovers. If you do not wish to share your juneberries with birds, net the plants.

## Selection of Plants

'Ballerina': The fruits are tender, sweet, and purplish black. The plant grows as an upright shrub or small tree, fifteen to twenty feet tall. The flowers are large, and the leaves turn purple-bronze in autumn. Hardy to Zone 4.

'Pembina': This saskatoon has large, sweet, full-flavored fruits. The large bush is upright growing and produces few suckers.

'Robin Hill': One of the earliest juneberries to bloom, this tree's flowers are pink in bud, then unfold to white. Hardy to Zone 4.

'Smoky': This saskatoon bears sweet, mild-flavored fruits. The bush produces many suckers and grows eight to ten feet high.

'Success': This variety, originating in the early 1800s, was the first juneberry variety to be selected for its fruit. The plant grows six feet high and four feet wide, bearing long clusters of fruits.

'Thiessen': This saskatoon bears large late-ripening fruits, with the best of flavors. The bush can grow fifteen feet high and wide.

# KIWIFRUIT
*(Actinidia* spp.)

**Pollination needs:** Except for a few self-fruitful varieties, kiwifruits need a separate male vine, which can pollinate about eight female, fruiting vines.

**Sun/shade requirements:** Full sun to partial shade

**Zones:** 7 to 9; hardy varieties, 3 to 7

Kiwifruits are johnny-come-latelies among cultivated fruits. Their emerald green flesh and pineapple-y flavor have captured the hearts of both fruit lovers and gardeners outside Asia in only the last century. The plants were initially introduced to the West from Asia primarily as ornamental vines. New Zealand growers were the first to cultivate the large fuzzy brown kiwifruit (*A. deliciosa*), now common in markets and grown in the United States primarily in California. The plants need a long growing season—200 to 225 days—to ripen their fruits and grow well only in Zones 7 to 9.

Although grown as ornamentals for more than a century, so-called hardy kiwifruits (*A. arguta* and *A. kolomikta*), have been "discovered" for their fruits in only the past decade or two. *A. arguta* and *A. kolomikta* are adapted from Zones 4 to 7 and 3 to 7, respectively, and will ripen their fruits in growing seasons only 130 to 150 days long. The fruits themselves have been overlooked because they are small and have smooth green skins that are hidden by, and blend in with, the foliage. But the skins on these grape-sized fruits are edible, and the flavor is a tad sweeter than that of

market kiwifruits. More recently, other species have been used in breeding, resulting in fruits with red flesh or skin.

*Hardy kiwifruits*

All the kiwifruits ripen in late summer or early autumn.

## Growing
Except for their ability to grow in colder climates, hardy kiwifruits can be grown in the same way as kiwifruits, so the directions that follow apply to both types.

Kiwifruits are prone to crown rot, so choose a site with perfectly drained soil, or else build up a mound of well-drained soil on which to plant the vines. Kiwifruits, even the hardy ones, are damaged by fluctuating temperatures while dormant, so keep the plants away from walls or other areas that tend to heat up periodically in winter.

The vines tolerate a wide range in soil pH, from 5.0 to 6.5, but are sensitive to soil salinity. Because of this sensitivity, divide applications of chemical fertilizers into two or three smaller doses, applied from spring until early summer. Alternatively, use a bulky organic fertilizer, such as compost or manure. In the West, avoid using feedlot manure because of its high salt concentration.

Kiwifruits are rampant vines incapable of holding themselves up, so you need to erect some sort of support for the plants. Let the plants ramble casually over a pergola or arbor or, for more serious fruit production, erect a T-trellis consisting of five wires strung between the tops of T's six feet high and five feet wide. In either case, allow about 200 square feet per vine, remembering that you need a separate male vine for pollination.

Space plants fifteen feet apart and train a single trunk from the ground up to the center wire of the trellis. When the developing trunk grows just above the height of the wire, cut it back to the wire and train shoots from the two topmost buds in opposite directions along the center wire. These two shoots will form permanent cordons, off which fruiting arms will grow, draping themselves over the outside wires. Each winter, shorten the developing cordon to leave two feet of the previous year's growth.

Because they grow so vigorously and bear fruit only on shoots growing off canes that grew the previous season, kiwifruit plants need annual pruning—ideally, twice a year. Winter pruning consists of shortening laterals growing off cordons (or growing off older laterals) to a length of about eighteen inches. After a lateral is three years old, cut it off completely to make room for a new young lateral. Also shorten cordons each winter so that they are no longer than seven feet. Summer prune by cutting back any overly rampant or tangled shoots and keeping the trunk clear of shoots.

Male plants are needed only for their flowers, so right after blossoming, prune flowering shoots back to new shoots. Prune heavily, removing about 70 percent of the previous year's growth.

## Pest Damage and Identification

Kiwifruits have few pest problems. The plants are among the many enjoyed by Japanese beetles. Cats sometimes like to scratch and gnaw on the plants, which have an effect similar to catnip. Protect young plants from cats, if necessary, with a cylinder of chicken wire. Mature plants do not need protection.

## Selection of Plants

'Abbott': Fruits are long and medium-sized. The vine has a medium to low chilling requirement and therefore is suitable for short winter areas, but it blossoms early.

'Ananasnaja' (also called 'Anna' or 'Ananasnaya'): This variety of hardy kiwifruit has relatively large and firm fruit, with excellent flavor. The vine is reliably productive.

'Blake': This female kiwifruit is self-fertile. The fruits are small and pointed and ripen about a month earlier than 'Hayward.' The vines are precocious and productive.

'Chico No. 3' (also called 'California Male'): This male kiwifruit plant blossoms midseason.

'Hayward' (also called 'California Chico'): This is the kiwifruit you usually find in markets. The fruits are large, tasty, and keep well. The vine has a high chilling requirement. Productivity and vigor are less than that of some other varieties.

'Issai': This variety of hardy kiwifruit is somewhat self-pollinating and commonly bears fruit the year after planting. The vine is less vigorous and somewhat less cold-hardy than that of most other hardy kiwifruit varieties. Pollination with a male increases fruit size and set.

'Matua': This male kiwifruit plant blossoms over a long period of time. The vine is cold-tender.

# MEDLAR
*(Mespilus germanica)*

**Pollination needs:**   Self-fruitful

**Sun/shade requirements:**   Full sun

**Zones:**   5 to 8

Medlar is related to apple, and the fruit is similar except that the medlar fruit is smaller and its calyx end—opposite the stem—is flared open. One other major difference: a medlar is rock hard when you harvest it in autumn and must be allowed to soften indoors before it can be eaten. Once softened, the inside of the fruit has the look and spicy taste of rich, old-fashioned applesauce.

Especially during the Middle Ages in Europe, medlar was valued for the beauty of the plant and the taste of the fruit. After the plant leafs out in spring, the tip of each short branch is decorated with a single white flower that resembles a wild rose.

## Growing
Medlar trees fruit best in full sunlight and well-drained soil. Medlars are grafted on pear, quince, hawthorn, juneberry, or seedling medlar rootstocks, so plant the tree with the graft union below ground level to induce the medlar scion to eventually form its own roots. Little pruning is needed beyond that necessary to train a young tree and to remove dead or crossing branches. The young plants are very precocious, so prune off the tips of short branches to remove some flower buds and channel more energy into growth.

## Pest Damage and Identification
Although medlar shares some pest problems common to apple, pear, and other relatives, these pests rarely cause enough damage to medlar to warrant control.

## Selection of Plants
'Dutch': This variety bears large fruits, two to three inches across. The tree is vigorous, with an almost weeping growth habit.

'Nottingham': This very old variety has small but tasty fruits. The tree has an upright growth habit.

*'Nottingham' medlar fruit*

'Royal': This variety has characteristics intermediate to 'Dutch' and 'Royal,' both in fruit size and tree form.

# MULBERRY
*(Morus spp.)*

**Pollination needs:** Most cultivated varieties are self-fruitful.

**Sun/shade requirements:** Full sun

**Zones:** 5 to 8

Wild mulberries abound over much of the country, but a backyard tree is more convenient and, if a selected variety, will have better-tasting fruits than its wild counterparts. There are three species of mulberry—black, red, and white—but the species names

74

bear little relation to the color of these blackberry-shaped fruits.

Best tasting of the mulberries is the black mulberry (*M. nigra*), native to western Asia and adapted to the dry, mild winter areas of Zone 7 on the West Coast. The red mulberry (*M. rubra*) is native to America but has naturally hybridized with white mulberry (*M. alba*), introduced from Asia over a century ago and now also wild here. Hardiness of these plants vary, with the hardiest ones adapted to Zones 5 to 8. With some varieties of mulberry, the ripening season can extend over a large part of the summer.

## Growing

The weedy nature of mulberries is testimonial to their tolerance for a wide range of site conditions. For best fruiting, plant a mulberry tree in full sunlight. Avoid planting mulberries near walkways or driveways, which will be stained by falling fruits.

Beyond planting and perhaps a little training, a mulberry plant needs no further care. You can train the plant as a large shrub with many stems or as a tree with a single trunk.

## Pest Damage and Identification

The major pest of mulberries is birds. Once a plant grows large, however, it bears enough fruits to satisfy both humans and birds.

## Selection of Plants

'Black Persian': A variety of black mulberry with delectable fruits, this is the one to grow if you live on the West Coast. This variety may go under other names.

'Illinois Everbearing': This variety is probably a hybrid of white and red mulberry and bears fruits that taste just about as good as those of 'Black Persian.' 'Illinois Everbearing' grows well in the East to Zone 5.

*Mulberry fruit on tree*

# NECTARINE

*(Prunus persica)*

**Pollination needs:**   Most varieties are self-fruitful.

**Sun/shade requirements:**   Full sun

**Zones:**   6 to 9

A nectarine is simply a peach without fuzz. Peach trees occasionally send out branches that yield fuzzless fruits, and nectarine trees occasionally bear fuzzy fruits.

## Growing

Refer to the plant portrait for peach, because the two fruits are grown in the same way.

## Pest Damage and Identification

Nectarine is threatened by the same pest problems as are peaches, so refer to the plant portrait about peach for this information. The smooth skin of a nectarine makes the fruit somewhat more susceptible to brown rot disease than is a peach.

## Selection of Plants

Although a nectarine tree never grows very large, consider planting a dwarf where space is limited. There are many genetic dwarfs, ranging in height from only a couple of feet to a few feet tall. For canning or cooking, plant one of the clingstone varieties, such as 'Red Gold,' 'Fantasia,' and 'Sparkling Red,' which tend to have firmer flesh. Freestone varieties generally have better quality for fresh eating.

*'Fantasia' nectarine*

'Fantasia': This freestone California nectarine has excellent flavor and is adapted even in the East.

'Hardired': This freestone nectarine, bred in Canada, has yellow flesh and good quality. The plant tolerates bacterial spot and brown rot, has large showy pink flowers, and is very hardy.

'Mericrest': This variety, bred in New Hampshire, is one of the hardiest nectarines. The small freestone fruit has excellent flavor and a dark red skin. The tree is medium large and productive.

'Snow Queen': This is a freestone nectarine with soft white flesh.

'Southern Belle': This large freestone nectarine has yellow flesh. The tree is productive and has a low chilling requirement. At full size, the tree is only five feet tall.

'Stark Gulf Pride': The ripe fruit of this freestone variety is very firm, with yellow flesh and excellent flavor. The tree has a very low chilling requirement.

# PAWPAW
*(Asimina triloba)*

**Pollination needs:**   Requires cross-pollination.

**Sun/shade requirements:**   Full sun

**Zones:**   5 to 9

Pawpaw is a native American fruit that hints of mango, papaya, and avocado, but mostly has the taste and texture of banana—hence, its other common names: Hoosier banana, Michigan banana, and poor man's banana. The tropical flavor does not keep you from being able to grow this fruit all the way from snowy Zone 5 to balmy Zone 9. The medium-sized tree even has a tropical appearance, with long, lance-shaped leaves that resemble those of avocado. Pawpaw fruits ripen in clusters—also very similar to bananas—in late summer or early autumn.

*Pawpaw fruits*

## Growing
Pawpaw is native to woodlands but will fruit best in full sunlight. However, very young plants like a bit of shade, which you can temporarily provide with an evergreen bough stuck in the ground on the south side of the plant. Pawpaws have long taproots, so plant either a potted tree or one that has been transplanted with care, in spring.

Pawpaws require little or no pruning once they have been trained to a sturdy framework of branches. If flowering becomes too sparse on a mature tree, head back some branches to stimulate growth, which will flower and fruit the next year.

Pawpaw flowers are not all that attractive to insects, so pollination sometimes is a problem on small plantings. Hand pollination is easy, though, and worth the effort because each flower can give rise to a cluster of fruits. Pick a few flowers off one tree, then dab the tip of an artist's brush back and forth from those flowers to attached flowers on another tree. Reverse the process to pollinate the first tree.

## Pest Damage and Identification
Pawpaw has no pest problems worth noting.

## Selection of Plants
Seedling pawpaws commonly bear good-tasting fruits, but to be assured of the highest quality and largest fruits, plant a named variety. Named varieties also bear their first fruits at an earlier age than do seedlings. Research is underway to find and develop new varieties

of pawpaw. Some of the best ones currently available include 'Fairchild' (early ripening), 'Overleese' (late ripening), and 'Sunflower' and 'Taylor' (late ripening and somewhat self-fertile).

# PEACH
*(Prunus persica)*

**Pollination needs:**   Most varieties are self-fertile.

**Sun/shade requirements:**   Full sun

**Zones:**   6 to 9

The best peaches—drippy, aromatic, and honey sweet—are those that you grow yourself. The trees also are worth growing because they never become very large and are very ornamental, with their showy pink blossoms and shiny, drooping leaves. Peaches are best adapted to Zones 6 to 9, but some especially hardy varieties can be grown in Zone 5.

*'Jimdandee' peach*

Although the botanical name implies origin in Persia, peaches actually are native to China. Long ago, peaches were carried westward, through Persia, then on to ancient Greece, then Rome. The Romans spread cultivation of the peach throughout their empire. Seeds were brought to America in the sixteenth century, where they found a welcome reception. Indians such as the Havasu in the Southwest soon began cultivating orchards of peaches, and in the Southeast peaches even formed wild stands ("Tennessee naturals").

Peaches grow readily from seed, and many seedlings bear high-quality fruits at a relatively young age. As a result, there are many, many varieties of peach, with new ones rapidly being developed.

## Growing

To keep your tree healthy, productive, and bearing the best-quality fruit, choose a site carefully. Peaches bloom very early in the season, so the site should not be one that warms up early in spring. Also critical to success with peaches is full sunlight and perfectly drained soil.

Train your young tree to a vase-shaped form, then prune every year. To promote rapid healing of pruning wounds and to more easily recognize and remove branches injured by winter cold, prune as late in the dormant season as possible—while the tree is in bloom is an ideal time.

Annual pruning is important for thinning out potential fruits and for stimulating growth of new wood, which is the only wood that bears fruit the following year. Use a combination of heading and thinning cuts to stimulate growth as well as to keep the tree low and open to light. The combined effects of pruning and fertilizing should stimulate three feet of new growth on young trees and eighteen inches of growth on mature trees. A mature peach tree needs severe annual pruning; remove enough wood each year so that a bird could fly right through your tree's canopy.

Even after you remove potential fruits by pruning, you probably will have to thin fruits after they set. While the fruits are still small, remove the excess so that those that remain are no closer than eight inches apart.

LORING

## *Pest Damage and Identification*

East of the Rocky Mountains, an insect called plum curculio damages developing fruits with a crescent- shaped scar. Injured fruits commonly drop or are susceptible to disease. Curculio is active during the six-week period beginning right after petal fall. Two or three sprays during this time should control the insect.

The oriental fruit moth is another insect pest that attacks peach fruits, leaving a large hole where it enters. This insect also bores into young stems, causing them to wilt at their tips. Besides spraying, this pest also has been controlled with pheromones (sold as Isomate-M, four tubes per tree, replaced after ninety days) and *Macrocentrus* parasite.

Catfacing of fruits can be a problem, especially in the Southeast. These sunken, corky lesions are caused by plant bugs such as the tarnished plant bug. The insects live among weeds and plant debris, so clean up areas adjacent to trees and also spray when necessary.

Fruits that become covered with a fuzzy gray fungus are infected with brown rot disease. Infected fruits darken, shrivel, harden, and often remain hanging in the tree. Clean up these mummies from the ground and the trees, because they are a source of the next year's infection. Spraying around bloom time and before fruits begin to ripen also helps control brown rot.

Dark spots pitting the surface of fruits are symptoms of bacterial spot disease. Besides spraying, your choice of varieties also can limit this disease. 'Redhaven,' 'Harrow Diamond,' and 'Harrow Beauty' show some resistance.

If the leaves on your peach tree are reddish, puckered, and curled, they have peach leaf curl disease. Again, resistant varieties ('Elberta,' 'Redhaven,' and 'Candor,' for example) and spraying help limit the disease. Cold winters kill the fungus.

When whole trees or large branches are wilting, suspect bark cankers or borers. Cankers are caused by a combination of factors, including wounds and winter cold damage resulting from lack of hardiness or late

*'Madison' peach*

fertilization. Borers are insects that generally attack weakened trees or trees that have already had their bark damaged, as by lawnmowers or cankers. To kill borers in the bark near ground level, sprinkle a ring of one ounce (less for young trees) of paradichloro-benzene mothballs around the base of the tree in autumn, before the soil cools to below 60° F. Mound soil up over the mothballs and against the trunk, leaving the mound in place for a month. By then the mothballs will have dissipated, and you can spread the soil back around the base of the tree.

## Selection of Plants

New peach varieties are constantly and rapidly being developed, improving pest resistance and flavor and extending the northern and southern limits for peach growing. Your first consideration should be choosing a variety adapted to your area in terms of tolerance for winter cold and chill requirement.

Consider also a variety's productivity and pest resistance. In some regions, such as California, pests are inconsequential; not so over most of the East. There are varieties that tolerate such diseases as brown rot and bacterial spot.

REDHAVEN

Southeast: 'Texstar,' 'TexRoyal,' 'LaSeliana'

West: 'Flavorcrest,' 'Suncrest,' 'O'Henry,' 'Fairtime'

# PEAR
*(Pyrus* spp.*)*

**Pollination needs:**   Requires cross-pollination.

**Sun/shade requirements:**   Full sun

**Zones:**   5 to 9

Also, of course, consider fruit quality. The best white-fleshed peaches taste as if they have been drenched in honey. Good varieties for the West include 'Babcock' and 'White Lady.' Few white-fleshed varieties are grown in the East; some varieties worth trying include 'Scarlet Pearl,' 'LaWhite,' and 'Morton.'

Yellow-fleshed peaches are either freestone or clingstone, depending on whether or not the pit adheres to the skin. Freestone peaches generally have high-quality, smooth, melting flesh. They also are easier to pit for canning or cooking. However, the firm flesh of clingstone peaches holds together better for cooking.

Some varieties of freestone (or almost freestone), yellow peaches recommended for different regions include (listed in approximate order of ripening):

Northeast: 'Harrow Diamond,' 'Redhaven,'
  'Jerseyqueen,' 'Harson,' 'Harrow Beauty,'
  'Encore,' 'Harcrest'

Two distinctly different types of pears are cultivated for their fruits. In Europe and North America, the major type is the soft, buttery European pear (*P. communis*). Throughout most of the Orient, the pear most commonly cultivated has been the Asian pear, also known as nashi (*P. pyrifolia, P. ussuriensis,* and *P. Bretschneideris*). Asian pears have sweet, crisp flesh whose cells burst with juice as you take each bite. Because of their firm flesh and usually round shape, Asian pears are sometimes known as apple pears.

Both Asian and European pears have been cultivated for thousands of years, and within each type there exists thousands of varieties. As the various species met in America in the nineteenth century, breeders hybridized them, infusing the European pear with the disease resistance or cold-hardiness of some of the Asian pears.

You can generally grow pears if you live in Zones 5 to 9. With some of the cold-hardier varieties, you can push pear growing into Zone 4 or even Zone 3.

## Growing

Plant pears in full sun and in soil that is at least reasonably well drained. Pears can tolerate wetter soils than can many other tree fruits.

Prune a pear tree every year, while the plant is dormant. On a young tree, do no more than the minimum amount of pruning necessary to train the plant to a central-leader form. During the training period, also bend branches down toward the horizontal to promote good form and to hasten early fruiting.

Once a tree reaches bearing age, prune lightly every year. Pear trees bear fruits on long-lived short spurs and therefore do not need to make much new growth every year to remain fruitful. The combined effects of fertilizer and pruning should stimulate about two feet of annual growth on a young tree or one foot of growth on a mature tree. As you prune, remove water sprouts, head back drooping stems, and cut diseased or damaged wood back into sound wood. On old trees, also thin out spurs where they are too crowded.

Each flower bud on a pear tree opens to a cluster of flowers, so pear trees can overbear. Thin fruits to about five inches apart.

## Pest Damage and Identification

In many gardens throughout the country, pears can be grown successfully without any attention to pest control. Elsewhere, a few pests threaten.

One of the most serious pests, where it occurs, is fire blight disease, which makes branches look as if they have been singed by fire. The blackened leaves remain attached to the stem, the tips of which curl over in a characteristic shepherd's crook. Because fire blight most readily attacks succulent growth, avoid excessive fertilizer or pruning, both of which can overstimulate a tree. One way to slow down an overly vigorous tree is to grow grass right up to its base to suck up extra water and fertilizer. (Do not do this with young trees, though, or you may stunt them.) Diligent pruning away of diseased wood can also control the disease. Sterilize pruning shears in alcohol between cuts in summer and keep an eye out in winter for dark, sunken fire blight cankers, where the disease-producing bacteria spend the winter. Varieties resistant to fire blight include European types such as 'Magness,' 'Maxine,' 'Moonglow,' and 'Seckel' and Asian types such as 'Shinko,' 'Seuri,' and 'Ya Li.'

*Fire blight*

Two other problems can be mistaken for fire blight. The first is *Pseudomonas* blight, a disease common in cool, moist autumn weather. Cankers of this disease do not ooze slime in spring, as do those of fire blight, but both diseases can be effectively controlled by pruning.

Blackened leaves also can be caused by superficial molds growing on honeydew dripped on pear leaves by an insect, the pear psylla, as it feeds. Differentiate between fire blight and this superficial mold by rubbing a leaf; if the black color wipes off, the problem is not fire blight.

Pear psylla can weaken pear trees, and can be controlled with dormant oil or insecticidal soap sprays.

A hole in a fruit and a fat white caterpillar within indicate codling moth damage. Reduce codling moth damage, when necessary, by thinning fruits (the insects prefer to enter fruits that are touching), by hanging pheromone traps in a tree, and by spraying.

## Selection of Plants

Full-size pear trees grow very large, but there are a number of semi-dwarfing and dwarfing rootstocks available. Quince is one such rootstock, producing a tree that eventually grows only about eight feet high.

A series of rootstocks, designated OH × F (they are hybrids of 'Old Home' and 'Farmingdale' pears), also offer various degrees of dwarfing. The OH × F series also are resistant to fire blight disease, which does not prevent a grafted scion from getting the disease but at least prevents the disease from traveling internally to kill the whole tree. The most dwarfing rootstock of the series, producing a tree comparable in size to that on quince rootstock, is OH × F 51. Moving up in size are OH × F 333 and OH × F 217.

The following is a selection of a few varieties from among the thousands of European and Asian pears:

'Bartlett': This common market pear, a European type, has a smooth yellow skin and is delicious when grown, picked, and ripened well. The fruit can be stored for a few months. The tree is productive and reliable. 'Bartlett' cannot pollinate 'Seckel.'

*'Bosc' and 'Anjou' pear fruits*

*'Comice' pear on tree*

*'Red Comice' and 'Green Comice' pear fruits*

'Beurre Bosc': This European pear has a russet skin, a crisp texture, and a sweet, rich flavor. The fruit keeps two to three months in storage. This variety is well adapted to northern and western regions.

'Comice': This European pear is considered by many to be the best flavored of all pears. The skin is greenish yellow, with a slight blush and russet. Store this fruit at least a month before ripening it for eating. Production is somewhat erratic.

'Hosui': This golden brown Asian pear has a sweet, rich flavor and stores well. The fruit is round. The tree has a somewhat weeping form.

'Kieffer': This variety is a hybrid of European and Asian pears, introduced in the nineteenth century. 'Kieffer' is suitable for cooking or canning but is poor quality for eating fresh. The fruits are large and yellow. This variety is adapted to many regions, including hot areas of the Southeast and lower Midwest.

'Magness': This European pear ranks very high for flavor and for resistance to fire blight. The fruit is sweet, juicy, and soft, with a greenish yellow skin having dark spots and light russet. 'Magness' is slow to come into bearing and cannot pollinate any other pears. This variety does well in the South and West.

'Nijisseikei' ('Twentieth Century'): This Asian pear is often seen in American markets. The fruit has an especially thin, tender skin, a sweet, juicy flesh, and stores well. This variety tends to bear crops in alternate years.

'Seckel' ('Sugar Pear'): Fruits of this European pear have a sweet, spicy flavor and a yellowish brown skin with a pale russet and a red cheek. This is the only European pear that you should not harvest until it has fully ripened on the tree. The tree is naturally small, resistant to fire blight, and suitable for growing almost everywhere. 'Seckel' cannot pollinate 'Bartlett.'

SECKEL

*American persimmons on leafless tree*

'Seuri': This large Asian pear is late ripening and has a flavor that hints of walnut. The tree is very productive, but the fruit does not store well.

'Ya Li' ('Duckbill Pear'): This old variety of Asian pear is pear shaped, with a sweet and aromatic flavor. The fruit stores well. 'Ya Li' is a good variety for warm regions.

# PERSIMMON
*(Diospyros spp.)*

**Pollination needs:**   Variable, depending on variety.

**Sun/shade requirements:**   Full to partial sun

**Zones:**   Oriental varieties, Zones 7 to 10; American varieties, to Zone 5

The botanical name for persimmon, *Diospyros*, translates as "food of the gods," referring to the honeylike sweetness of the fruits. The Oriental persimmon (*D. kaki*), also known as kaki, long popular in the Orient, looks like a tomato and has flesh that is either crisp or like jelly, depending on the variety. The fruit of the American persimmon (*D. virginiana*) is smaller than that of the Oriental persimmon, with a drier, richer flesh, almost like a dried apricot. Most Oriental persimmons can be grown in Zones 7 to 10. American persimmons are cold-hardy to Zone 5.

Both persimmon species grow into full-sized trees that are either male or female. Female, fruiting trees may or may not need cross-pollination. Most Oriental persimmons do not need pollination to set fruit, but some varieties can be eaten when crisp only if they have been pollinated. Most, but not all, varieties of American persimmon need pollination. Fruits of persimmons that set fruit whether or not they are pollinated are still affected by pollination, with unpollinated fruits being smaller and lacking seeds.

## Growing

Persimmons are not especially finicky about site conditions. The trees tolerate most soils, except those that are waterlogged, and either full sun or a little shade. Persimmons have long taproots, so plant either a potted tree or one that has been transplanted with care, in spring.

Once a persimmon tree has been trained to a sturdy framework of branches, little further pruning is necessary. Aside from removing misplaced, dead, or diseased branches, prune each winter to stimulate some new growth, on which fruit will be borne the following season. Also head back some young wood to reduce the crop load and keep bearing limbs in near the trunk. If fruit set is very heavy, especially with Oriental persimmons, thin the fruits by hand after they have set.

Harvest American persimmons and so-called astringent varieties of Oriental persimmons when they are fully colored and very soft. Pick the fruits of Oriental persimmon varieties that are edible while firm as soon as they are fully colored. Oriental persimmons and, to a much lesser extent, American persimmons can be picked underripe and ripened indoors. You can speed this ripening process by putting near-ripe fruits into a plastic bag with an apple.

## Pest Damage and Identification

A persimmon girdler and a persimmon borer sometimes damage the bark on trees, but persimmons generally are free of pest problems.

## Selection of Plants

The most important factor in selecting a persimmon tree is to choose a variety adapted to your climate so that the tree will survive your winters and have a sufficiently long season in which to ripen its fruits. In the coldest areas, grow American persimmons, making sure that the varieties you plant will ripen within your season. In warmer regions, you can grow astringent Oriental persimmons and, in still warmer areas, non-astringent Oriental persimmons. Where pollination is needed, you will have to either plant a separate (nonfruiting) male tree or graft a male branch onto your female tree. American persimmons grow wild throughout much of eastern United States, and wild trees can often meet your tree's pollination needs.

'Early Golden': This variety, found growing wild on an Illinois farm in 1880, was the first American persimmon to be named. The fruits have good size and quality. For best yields, provide a pollinator.

'Eureka': One of the hardiest Oriental persimmons, this variety has delicious red fruits. 'Eureka' needs pollination to set fruit.

'Fuyu' ('Fuyugaki'): This Oriental persimmon is the one you sometimes see in markets. You can eat the large orangish red fruits while they are still firm. 'Fuyu' does not need pollination but yields more consistently if pollinated.

'Hachiya': This Oriental persimmon sometimes turns up in our markets. The fruits are large, orangish red, and cone shaped, and have excellent flavor. 'Hachiya' does not need pollination and should be soft before you eat it.

'Meader': This American persimmon, adapted to cold regions, usually does not need a separate pollinator. The fruit ripens early but loses its astringency slowly.

'Pieper': This variety of American persimmon is suitable for cold regions. Pollination is not necessary to produce fruits, which are small and reddish yellow.

# PLUM
*(Prunus spp.)*

**Pollination needs:** Damson plums and most European plums are self-fertile; Oriental and American hybrid plums require cross-pollination.

**Sun/shade requirements:** Full sun

**Zones:** European varieties, Zones 5 to 8; Damson, Zones 5 to 9; Oriental varieties, 6 to 10

With so many species of plums, it is no wonder that there are so many types of plum fruits. Their flavors range from sweet to tangy, and their colors from yellow to blue to almost black. Some plums are no bigger than cherries, while others are as large as apples. Plum is a "stone fruit," closely related to peach, nectarine, apricot, and cherry, and has even been hybridized with

*Bowl of many varieties of plums*

some of these relatives. Plum-apricot hybrids are known as pluots or apriums, depending on the amount of plum or apricot in a particular variety's parentage.

Four types of plums are grown for their fruits. European plums (*P. domestica*) come in various colors and shapes and are suitable for growing in Zones 5 to 8. These fruits are good for eating, cooking, and if they are prune plums, for drying. Damson plum (*P. institia*) is another plum from Europe, also adapted from Zones 5 to 9. Damson plums are usually used in cooking, for making preserves, jams, and pies. Oriental plums (*P. salicina*) are large fruits adapted from Zones 6 to 10. Hybrid plums are combinations of native American plums, of which there are numerous species, with Oriental plums. Plants of hybrid plums are generally bushy and very cold-hardy, and the fruits vary in quality from those suitable only for cooking to those that taste good fresh.

## Growing

All plums require a site that has well-drained soil, full sunlight, and is free from late spring frosts. Space plants according to their eventual size, which ranges from fifteen feet high and wide for an Oriental plum to only six feet high and wide for some hybrid plums.

Train your young plant to an open-center form, except in the case of some upright-growing varieties, which naturally take on the form of a central-leader. Plums bear fruit both on the previous season's shoots and on short-lived spurs on older wood and therefore need only a moderate amount of annual pruning to maintain a slow but steady supply of new bearing wood. Pruning and fertilization combined should stimulate about three feet of branch growth each

season on a young tree, or about eighteen inches of growth on a mature tree. Prune also to keep the plant open to light and air and to remove diseased or damaged wood.

In most cases, you do not have to thin plum fruits. However, the large fruits of Oriental plums can weigh down a branch to the point of breaking, so thin these fruits to five inches apart if the crop is heavy. You also can prevent limb breakage by shortening long branches loaded with fruits.

*Brown rot of plum fruit*

## Pest Damage and Identification

East of the Rocky Mountains, plum curculios commonly lay their eggs on developing fruits, leaving crescent-shaped scars and usually causing the fruits to drop. Fruits that do not drop are ruined, either from feeding of the larvae or from diseases entering through the scar. Spraying during the six weeks following bloom is the best method for controlling curculios.

Plums that become covered with a fuzzy gray fungus, then darken and shrivel up to mummies, have brown rot disease, prevalent in humid weather. In addition to spraying, you can limit this disease by planting a variety with some resistance and by picking off mummies, which provide inoculum for next year's infection.

Brown sunken pits in fruits are symptoms of bacterial leaf spot, a disease common in the Southeast. On the leaves this disease causes angular dark spots. The best control for this disease is to plant a resistant variety such as 'AU-Rosa,' 'AU-Roadside,' 'AU-Amber,' 'Green Gage,' or 'Ozark Premier.'

Also prevalent in the Southeast is plum leaf scald disease, which stunts or kills whole trees. There is no cure, so dig up infected trees and plant a more resistant variety such as 'AU-Rosa' or 'AU-Amber.'

A thick, tarry coating on branches indicates black knot disease. Pruning off diseased portions of the plant helps control the problem. Resistant varieties include 'AU-Rosa,' 'AU-Roadside,' 'AU-Amber,' 'Crimson,' 'Santa Rosa,' 'Fellenberg,' and 'Shiro.'

*Black knot disease on plum*

ONEIDA

ALDERMAN

## Selection of Plants

Dwarfing rootstocks are available for plums, but even standards do not grow into very large trees. Choose a special rootstock if you have a special soil problem— 'Nemaguard' peach rootstock for nematode problems, for example.

Refer to nursery catalogs for the specific pollination needs of hybrid plums.

'AU-Roadside': One of the 'AU' series of plums developed in Alabama for growing in the Southeast. The fruits are soft and dark red, with excellent flavor. The plant is very resistant to disease.

'Bluefre': Fruit of this variety of European prune plum is large, blue, and firm. The fruits hang well on the tree even after they are ripe. The tree has an upright growth habit.

'Green Gage' ('Reine Claude'): This old variety of European plum, named after the wife of Francis I of France, has heart-shaped fruit with a green skin and a rich, sweet, juicy, amber flesh. The tree is productive.

'Monitor': This hybrid plum has medium-large bronze fruits of good quality. Rain during ripening may cause the fruit to crack. The plant is very cold-hardy, productive, and vigorous.

'Mount Royal': Fruits of this European variety of prune plum are dark blue and medium-sized. The plant is very cold-hardy.

'Ozark Premier': This Oriental plum is adapted to growing in the Southeast. The fruits are very large and red, with a juicy, tart, yellow flesh. The plant is hardy and productive.

'Pipestone': This hardy hybrid plum bears large fruit having a red skin and yellow flesh. The skin is tough but peels easily. The flesh is somewhat stringy but has excellent flavor. This variety is hardy and reliable but pollen sterile and therefore cannot pollinate itself or other plums.

'Stanley': This European type of prune plum is similar to 'Italian' (a variety of prune plum commonly sold in markets) but ripens earlier and is more productive and widely adapted. The fruit is large, with a blue skin and yellow flesh.

GREEN GAGE

STANLEY

'Superior': A hybrid of Oriental and American plums, this variety bears early large red fruits having good flavor. Fruits of this variety require thinning. Plants are vigorous and productive.

# POMEGRANATE
*(Punica granatum)*

**Pollination needs:** Self-fruitful

**Sun/shade requirements:** Full sun

**Zones:** 9 to 10

Pomegranate is an ancient fruit, originating in the Mideast, and best adapted to growing in climates having cool winters (hardy to Zone 8) and hot, dry summers. The seedy fruits are delicious fresh, or you can squeeze juice from them to make a refreshing syrup or drink.

Do not overlook the ornamental value of this plant. The small multistemmed tree or large shrub becomes covered with flamboyant orangish red flowers throughout spring and sometimes on through summer. The plants are semi-evergreen in regions where winters are mild.

## Growing

Pomegranate plants thrive in heat and sun but are not finicky about soil. For extra heat in less-than-ideal climates, grow pomegranate near a south-facing wall.

Train your plant to either a single trunk or to five or six trunks. A plant with multiple trunks requires less frequent pruning when young and suffers less when cold causes some injury.

Once your plant is shaped, prune it lightly every year to maintain production. Flowers and fruits are borne on short spurs on two- and three-year-old wood growing mostly at the outer edge of the plant. Light pruning encourages growth of new wood to replace wood that is no longer productive and also thins out excess fruits so that those that remain grow larger.

Pick pomegranates by clipping the fruits off after they have fully colored. Do not let ripe fruits hang on the plant for too long or they may split, especially following rain.

## Pest Damage and Identification

Backyard pomegranate plants rarely have pest problems.

## Selection of Plants

As expected of such an ancient fruit, many varieties exist, varying most notably in seed size and hardness and in skin color. Varieties with light-colored skin are usually sweeter than dark varieties. The most widely available variety is 'Wonderful,' with deep crimson skin and relatively small tender seeds. 'Spanish Sweet' ('Papershell') is a popular dooryard tree with large pale yellow and pink fruits and sweet kernels.

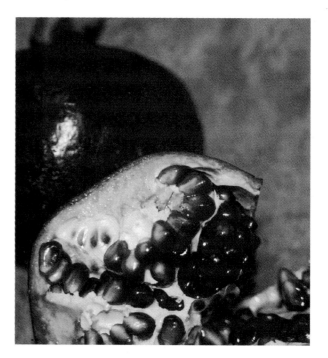

*Pomegranate fruit*

# QUINCE
*(Cydonia oblonga)*

**Pollination needs:** Self-fruitful

**Sun/shade requirements:** Full sun

**Zones:** 5 to 9

*Quince fruit*

A quince fruit is shaped something like a cross between a pear and an apple and has a fuzzy yellow skin. Although inedible raw, a bit of quince adds pizzazz to an apple pie or applesauce or cooks into its own delectable jelly. The fruits are borne on a small, sometimes multistemmed tree having large white flowers and—with age—picturesque gnarled branches.

Other "quinces," such as the flowering quince (*Chaenomeles* spp.), are grown for their showy flowers. Their fruits are edible but not very palatable.

## Growing

Give quince a site in full sun, with well-drained, moderately fertile soil.

Quince grows either as a shrub or small tree. In either case, shape the plant in its youth with a permanent framework of spreading limbs that will be open to sunlight and air. Once the plant has been shaped, prune it lightly every year to keep the center open and to remove dead and diseased wood.

## Pest Damage and Identification

Quince is closely related to apple and pear and is potentially affected by many of the same pests, most notably fire blight and rust diseases. Rarely, however, are pest problems serious enough to warrant action.

## Selection of Plants

'Pineapple': This variety, developed by Luther Burbank, bears large, round, golden fruits having a flavor hinting of pineapple. The white flowers are tinged with pink.

'Smyrna': This old variety, from Turkey, bears very large pear-shaped fruits with a golden yellow skin and a light yellow flesh.

*Quince tree at the Cloisters, New York City*

# RASPBERRY
*(Rubus spp.)*

**Pollination needs:**   Self-fruitful

**Sun/shade requirements:**   Full to partial sun

**Zones:**   3 to 8

Raspberries come in a variety of colors—red, yellow, purple, and black—each with its own distinctive flavor. Purple raspberries are hybrids of red and black raspberries and, like black raspberries, tend to have firm fruits that are more tangy and seedy than those of red raspberries. Incidentally, you can tell a black raspberry from a blackberry because the core stays on the plant when you pick a ripe black raspberry but stays in the fruit with a blackberry.

Growth habits vary among raspberries. Red and yellow raspberries grow on upright nonbranching canes that are either prickly or smooth. As the roots of red and yellow raspberry plants spread, the buds on those roots grow into new canes. Purple and black raspberries have arching canes and side branches, all armed with stiff thorns. These plants can "hopscotch" along as the tips of their canes touch the ground and root to make new plants.

All raspberries have perennial roots and biennial canes. In most cases, the canes just grow their first year, then bear fruit and die their second year. You get an annual supply of fruit, however, because new canes are growing at the same time older canes are fruiting.

The canes of one type of raspberry, so-called everbearing types, begin to bear fruit near their tops toward the close of their first season, then finish fruiting lower down on these same canes earlier in their second season. The result: two crops per season, a summer crop on second-year canes and a late summer or fall crop on first-year canes.

The most cold-hardy are the red and yellow raspberries, followed by purple raspberries, then black raspberries. A relatively new red raspberry adapted to very warm climates is the 'Bababerry.'

## Growing
Raspberries are relatively easy to grow, provided you pay attention to site selection and pruning. The plants tolerate a little shade but fruit best in full sunlight. The soil should be well drained, slightly acidic, and rich in organic matter. Avoid disease problems by not planting near wild raspberries or blackberries or in ground where you recently grew these plants or tomatoes, peppers, eggplants, or potatoes. In general, purple and black raspberries are more prone to diseases than are red and yellow raspberries.

Most raspberries are easier to manage if you trellis them. Sink a sturdy post into the ground at each end of the row, then string two or three wires (12 gauge) between the posts, with the top wire five feet above ground and the others lower down. When set three feet apart in rows six feet apart, red and yellow raspberry plants eventually fill in the row to make a solid swath of plants. Plant black or purple raspberries three feet apart in rows eight feet apart. No trellising is

*Raspberry hedgerow before pruning*

*Raspberry hedgerow after pruning*

necessary with any red or yellow raspberry that has stiff, upright canes, such as 'Heritage,' or with black or purple raspberries trained low.

Another way to grow raspberries is in the hill system. Space plants five to eight feet apart in all directions, with a sturdy post on which to tie the plants at each hill. Do not allow the plants to spread.

Annual pruning is a must to keep raspberries from becoming overgrown, diseased, and difficult to pick. Prune red and yellow raspberries in three easy steps in late winter. First, cut away at their bases all canes that fruited the previous summer. (You can also do this right after they finish fruiting.) Next, thin out new canes, leaving those that are thickest and healthiest. In a hedgerow, keep the row no wider than twelve inches and canes no closer than six inches apart; in hills, remove all but six new canes. In each case, always remove the thinnest and diseased canes first. Finally, top the canes to a height of about six feet.

In the case of everbearing raspberries, follow the same steps, but top the remaining canes to below the height where they finished fruiting the previous fall. You will see old fruit stalks still dangling from the canes. An easier way to prune everbearers is just to mow the whole planting to the ground each fall. This sacrifices the summer crop but helps prevent problems with diseases, winter kill, and deer damage.

Follow the same steps for pruning black and purple raspberries as described for red and yellow raspberries, with two additional steps. The first additional step is to pinch out the growing tips of new canes during the summer, when they are two to three feet high. Not all canes reach this height at the same time, so go over the planting repeatedly, pinching as needed. Pinching stimulates growth of branches, on which fruits are borne the following year. The other additional step is to shorten each of those branches in winter to about twelve inches.

## Pest Damage and Identification

Regular pruning thwarts many pest problems by removing pest-damaged and excess canes. Discolored spots or lesions on canes are symptoms of cane diseases, to which black raspberries generally are most susceptible. Remove infected canes as you prune. A cane that partially dies back may have a resident cane borer, which you can kill by cutting off and destroying the cane six inches below the point of entry (marked by a swelling). Dig out and destroy whole plants of black or purple raspberries if they show symptoms of orange rust—bright orange pustules on the undersides of the leaves.

Dieback of whole canes can be traced to a number of causes, including verticillium wilt, crown gall, root rot, or cane borers. Dig out and destroy such plants. Avoid such problems in the future by planting disease-free plants in soil that is also free from disease—where raspberries or other hosts to their diseases have not recently grown.

Eventually, a raspberry planting will decline as pests make inroads. Plan on setting out a new planting—at a new location and with new plants—after a decade or less.

## Selection of Plants

'Allen': Fruits of this black raspberry are firm, juicy, and sweet. The plants are fairly disease-resistant and ripen their crop over a short period. This variety is well adapted to the northeastern and north-central states.

'Bababerry': This everbearing red raspberry originated in southern California and is adapted to regions with hot summers and mild winters. The canes are sprawling and very vigorous. The fruits are large, tangy, and firm.

'Boyne': This red raspberry is notable for being very cold-hardy and productive. The canes are upright and the fruits are dark red and fairly sweet.

'Brandywine': This variety yields very large, purple, tangy raspberries. The plant is vigorous and semi-erect and does not spread by suckers.

'Bristol': This black raspberry bears large, black, sweet fruits. The plants are upright.

'Canby': This is a thornless red raspberry with bright red fruits. The sprawling plants are vigorous and resistant to virus diseases. 'Canby' enjoys cool summers and therefore does well in the Northwest.

'Fallgold': This yellow everbearing raspberry yields very sweet fruits. The plants are productive, upright, and widely adapted.

*'Fallgold' raspberry fruits*

'Heritage': Fruits of this red everbearing raspberry have fair flavor. The plants are upright and widely adapted.

'Latham': This older variety of red raspberry yields large deep red fruits with good flavor. The plants are productive, disease-resistant, and very cold-hardy.

98

'Meeker': This variety yields large, red, sweet fruits and is well adapted to the Northwest (replacing 'Willamette'). The plants are not very cold-hardy but are productive, yielding their fruits over a long season on long, willowy canes.

'Munger': This black raspberry yields fruits that are large, plump, firm, and not very seedy. The plants are somewhat disease-resistant, and do especially well in the Northwest.

'Royalty': Fruits of this purple raspberry are very large and tangy, although sweeter than 'Brandywine.'

'Taylor': The fruits are large, with excellent flavor. The canes are sturdy, vigorous, and productive.

# STRAWBERRY
*(Fragaria spp.)*

**Pollination needs:**   Self-fruitful

**Sun/shade requirements:**   Full sun

**Zones:**   3 to 10

"Doubtless God could have made a better berry, but doubtless God never did," wrote Izaak Walton, an English author of the seventeenth century, about the strawberry. And Walton had not experienced the large sweet garden strawberry (*F.* × *ananassa*) as we know it today. Walton's berries were probably alpine strawberries (*F. vesca*) and musk strawberries (*F. moschata*), occasionally still cultivated today for their smaller but

*'Taylor' raspberry fruits*

*Alpine strawberries*

highly aromatic fruits. Today's garden strawberry arrived on the scene in only the last two hundred years, the result of a chance hybridization—in Europe—of two American species, one from our East Coast and the other from our West Coast.

Today there are three kinds of garden strawberries. The first type is the June-bearing strawberry, yielding one crop per season, in spring or early summer. So-called everbearing strawberries were developed early in the twentieth century. This type of strawberry actually bears two crops per season, the first in spring and the second in autumn. The third type of garden strawberry, the day-neutral strawberry, was introduced about 1980. While other types of strawberries flower and fruit in response to day length, day-neutrals, as their name implies, are unaffected by day length and can flower and fruit all summer long. Day-neutrals are truly everbearing and usually are listed in nursery catalogs along with the older type of everbearing (two-crop) strawberries.

With the many varieties of strawberries available, you can grow this luscious fruit almost anywhere.

## Growing

Although strawberries will fruit over a broad range of climates, they will not perform well without full sunlight, good soil, and ample water. "Good soil" for a strawberry plant is fertile, rich in organic matter, and slightly acidic. Before you plant, fertilize the bed, dig in plenty of organic matter, and get rid of weeds, especially perennial weeds. If drainage is less than perfect, grow plants in raised beds that are six inches high and two to three feet wide.

With time, strawberries pick up diseases from wild plants, so have a new bed ready to fruit before you dig up your old bed. To delay the onset of disease problems, replant new beds with disease-free plants from a nursery rather than extra plants from a generous neighbor. Also, choose a site where strawberries or other plants (tomato, pepper, eggplant, potato, and raspberry) having pests in common with strawberries have not recently grown.

Choose from one of three methods of laying out a strawberry bed. The first method is to plant a matted row, in which you set plants far apart, at a two-by-four-foot spacing, then let runners (new plants that form along trailing stems) fill in the bare spaces between the mother plants. Keep the matted row that develops no wider than eighteen inches by cutting off runners that grow beyond that limit. The second planting method, the hill system, yields a larger first harvest but requires more initial maintenance and plants. To plant in hills, set a double row with one foot between plants and one foot between the rows. Pinch off all runners that form. If you plant more than one double row, allow four feet from one double row to the next. The hill system is best suited to alpine strawberries, day-neutral strawberries, and June bearers such as 'Darrow' and 'Canoga,' all of which make few or no runners.

A third option in planting strawberries is the spaced-plant system, a combination of the first two. Set plants two feet apart in rows three feet apart and allow only four well-placed runners to form around each mother plant.

Maintain a year-round mulch around your strawberry plants. Aside from the usual benefits, mulches

*Netted strawberries*

also keep ripening strawberry fruits clean of soil. Where winter temperatures drop below the teens, cover the whole planting with a fluffy mulch for cold protection and to prevent frost from heaving plants out of the soil. Wait to cover the tops of the plants with winter mulch until the soil has frozen to a depth of one inch. Straw is an ideal mulch (and provides one explanation for the name *strawberry*), but pine needles and other nonmatting leaves, such as those of oak, also are suitable.

After a couple of years, a strawberry bed will need annual renovation. Renovation prevents overcrowding, limits diseases and weeds, and maintains a population of young, fruitful plants. Renovate immediately after harvest, beginning by cutting off all the leaves with hand clippers or a mower set high, then fertilizing. Dig out old plants and thin out young plants so that each plant has a square foot of space. Mulch and water the bed, and the plants will soon start growing again.

Do not renovate beds of everbearing and day-neutral strawberries. Just replant them when they grow old and nonproductive.

Be alert for spring frost when your strawberries are flowering. If frost threatens, pull mulch back over the plants or throw a blanket over the bed.

## *Pest Damage and Identification*

With correct fertilization and annual renovation, the major pest your strawberry bed will experience will be birds. The only sure way to thwart them is to cover the bed with netting, held above plants so that birds cannot peck through at the fruits. In moist weather, fruits also may be attacked by slugs, which work at night but leave telltale shiny trails. Trap slugs in shallow pans of beer or block them with barriers of copper flashing.

In wet weather you also may find rotting fruits. Mulch and adequate plant spacing lessen this problem. When you harvest berries, pick off rotten ones and toss them into a separate container for disposal, to prevent further spread of disease.

If whole plants decline, suspect root problems. Blackened roots or rat-tailed roots (no side roots) are problems associated with poor drainage, in which case replant at a better drained site, using new plants. Wilting plants with no evidence of root damage may have verticillium disease, for which there is no cure except to replant at a clean site or with a verticillium-resistant variety, such as 'Catskill,' 'Sunrise,' or 'Surecrop.' Many plants, including raspberry, tomato, pepper, eggplant, and potato, can also host verticillium, so avoid planting where these plants or strawberries recently grew.

*'Earliglow' strawberry fruits*

## Selection of Plants

New strawberry varieties are constantly being developed, and there are varieties specifically suited to certain regions. Contact your local Cooperative Extension for the names of varieties best suited for your garden. The following is a selection of some of the more popular varieties in each region.

**North:**

'Catskill': Berries of this variety are very large, soft, and bright crimson, ripening midseason. The plants are adaptable and produce many runners.

'Guardian': The fruits are large, firm, glossy, and light red, with good flavor. The plants runner prolifically.

'Ozark Beauty': This everbearing variety bears two crops per season of large, sweet, bright red fruits.

'Raritan': The medium-sized fruits ripen early and are firm, with good flavor. Plants produce many runners.

'Sparkle': The fruits are medium-sized and soft, with very good flavor. Plants produce many runners.

'Tristar': This day-neutral variety produces fruits with excellent flavor. The plants are disease-resistant.

**South:**

'Florida Ninety': The berries of this variety are long, soft, and good quality. The plants produce many runners.

'Tangi': The berries are medium large and moderately firm, with good flavor.

**Central:**

'Blakemore': The berries are small, bright red, and tart. The fruits ripen early in the season on disease-resistant plants.

'Surecrop': This variety bears large glossy red fruits early in the season. The fruits have good flavor. The plants produce many runners and are disease-resistant.

'Tennessee Beauty': The berries are medium-sized, glossy red, and ripen in midseason with good flavor. The plants produce many runners and are disease-resistant.

**West:**

'Brighton': A day-neutral variety that yields very large juicy berries having good flavor. The plants are very productive and produce few runners.

'Hood': The berries are large and moderately firm, with a mildly tart flavor. The plant is disease-resistant but not very winter-hardy.

'Quinault': This everbearing variety yields soft berries with good flavor. The fruits are extremely large, good for cooking and eating fresh but not for freezing. The plants produce many runners.

'Tioga': The berries are large and very firm, with good flavor. The plants are productive and resistant to some diseases.

'Totem': This variety yields moderately large fruits having excellent flavor. The fruits ripen early. The plants produce many runners.

# Chapter 4

# PESTS

*A*LL *sorts of creatures, both large and small, may vie for your luscious fruits. If you start with clean, healthy plants, then keep them healthy with correct siting, pruning, watering, and feeding, you can avoid most pest problems—all pest problems, in many cases.*

There may be more than one way to control a pest when necessary. Visual traps, or those baited with sex attractants, often are effective at drawing pests away from fruits or preventing mating (and hence, laying of fertile eggs). Physical barriers prevent pests from getting to plants or fruits. Consider planting varieties resistant to pests that cause problems.

*Bagging grapes*

# Spraying

Spraying is just one of the many ways to control a pest. The spray itself may consist of something as benign as a forcible stream of water or a bacterium (such as *Bacillus thurengiensis*, marketed under such trade names as Dipel and Thuricide) that specifically infects grape berry moth and related caterpillars. At times you may have to resort to sprays of more potent pesticides.

There are two approaches to preventing pest problems with pesticide sprays. The first approach is preventative, whereby you apply a multipurpose fruit spray—containing materials to control major disease and insect pests—on a regular schedule. The second approach requires that you pay close attention to your plants and, when a pest problem appears, apply a pesticide at the optimum time, targeted specifically at that pest. The latter approach results in less spraying.

In either case, read the pesticide label to make sure the particular pesticide is effective against the particular pest(s) on the plant you intend to spray, then follow the label directions carefully. Pesticides are poisons, and by following directions explicitly, you minimize potential harm to both you and the environment while maximizing a material's effectiveness. Furthermore, if you do not use a pesticide according to directions on the label, you are breaking the law!

Major pests specific to each fruit are listed in the plant portrait section in chapter 3, but a few others are more general feeders.

*Apple scab on fruit*

*Rabbit damage to apple tree*

Aphids, for example, are small, soft-bodied insects that suck sap from a wide variety of plants. Look for them on the undersides of leaves. Where aphids become a problem, try hosing them off with plain or soapy water or prevent ants from "herding" aphids by putting a band of masking tape coated with sticky Tangletrap around the trunk of a tree.

Scale insects look like small white or gray bumps on the bark (not to be confused with the bark's natural openings, or lenticels). Scale insects also suck sap from plants and cause spots on fruits. Control scale, if necessary, with a dormant oil spray just before growth begins in spring.

Larger animals to watch out for include birds, rodents, and deer. Scare tactics—pie pans, reflective tape, and colored balloons—sometimes work against birds, but the surest preventative is netting. Black cotton thread is somewhat effective if you drape it throughout a tree by letting a spool of it unroll as you throw it back and forth among a tree's limbs.

*Young tree protected for winter with soap and paint*

*Hardware cloth and paint protecting mature tree for winter*

*Mouse girdling*

Prevent rodents from gnawing at tree bark in winter by enclosing the lower trunk in a cylinder of quarter-inch mesh hardware cloth or a plastic protector made for this purpose or by wrapping the lower trunk with paper tree wrap. Protection should extend an inch or so into the soil. Remove protectors, except hardware cloth, in spring so as not to leave places in which insects can hide.

Prevent deer from eating the stems of trees by spraying or festooning the branches with commercial repellents or with homemade repellents such as deodorant soap, bags of hair, or mixtures of egg and hot pepper. Fencing can also be effective, but its effectiveness depends on the hunger of the deer and the height of the fence.

## TWO FINAL THOUGHTS

Learn to tolerate a certain amount of damage and fruit loss. The extra spraying or other effort needed for maximum yields of 100 percent blemish-free fruits is not justified in a backyard. And anyway, many blemishes are no more than skin deep. Being able to set your own standards is one of the "luxuries" of growing your own fruits.

Perhaps most important is to keep a close eye on your plants, not only to watch for pests but also to observe how the plant is growing as you prune, feed, and water. Your close attention will increase your personal satisfaction in growing fruits, as well as bring in the bounty. As an old saying goes, the best fertilizer is the gardener's shadow.

# Chapter 5

# HARVESTING AND STORING YOUR BOUNTY

*J*UST *as you would not bother to cut a wilted, browning, old rose blossom to put into a vase, similarly you should not harvest fruits except at their prime. After all, one reason you choose to grow fruits in your backyard is so that you can savor them at their luscious best.*

With few exceptions, harvest fruits when they are fully ripe. Most fruits do not improve in flavor if they are picked underripe, then left sitting on a kitchen counter. Softening does occur, but such softening in many cases is more akin to incipient rotting than to ripening.

The most obvious way a fruit tells you that it is ripe is by changing color—ripe strawberries are red, ripe apricots are orange, and ripe blueberries are blue. Do not even consider picking a fruit until it is fully colored. With some fruits this color change is subtle. Red apples and peaches do not necessarily turn red when ripe, because direct light is needed for this color change. In shaded portions of a tree, these fruits indicate their ripeness when the background color changes from grass green to very light green or creamy yellow. On the other hand, ripe blueberries, cherries, grapes, and plums color up whether or not exposed to direct light.

When fully ripe, a fruit also softens a bit and separates readily from the plant. In the evolutionary scheme of things, a fruit is merely a vehicle to spread seeds, and a mature fruit with fully developed seeds falls easily from the plant. For harvesting mulberries in quantity, spread a sheet on the ground beneath the tree and give the limbs a shake. Ripe pawpaws and American persimmons are delicious by the time they hit the ground. Blanket the ground beneath these trees with a soft mulch, such as straw, to soften the impact of the ripe fruits when they fall.

Waiting for ripe fruit to fall is not a practical way to harvest most other fruits and often occurs when fruit is past its prime for eating. Pick most tree fruits by cupping them in the palm of your hand, then giving a slight twist around, then up. If the fruit is ripe, it will part readily from the plant.

Don't rush the harvest, especially when harvesting berries. Blueberries, for example, color up days before the fruits are truly ripe. To separate the blue ripe blueberries from the blue unripe blueberries, cup your hand beneath a cluster of fruits, then tickle them. Only the ripe fruits will fall off into your waiting hand. Some gooseberries—'Oregon Champion,' for example—are as green when they are ripe and sweet as when they are underripe and sour; tickle these also, gingerly because of the plant's thorns. Blackberries remain quite puckery until they are so soft that they stain your fingers as the fruits practically drop off the plant into your hands.

There are exceptions to the rule of waiting until fruits are fully ripe before harvesting them. For cooking into jams and pies, fruits hold together better and provide a welcome tangy flavor if picked slightly underripe. Fruits that you intend to store keep better if harvested before they are fully ripe. Some, such as apples, kiwifruits, and persimmons, can actually ripen during storage. Late apples such as 'Idared' and 'Newtown Pippin' improve in flavor during their storage period.

Finally, there is the case of European pears (with the exception of 'Seckel') and medlars, both of which *must* be picked underripe to ripen off the plants. Left on the tree until fully ripe, a European pear will be brown and mushy inside. (Pick Asian pears when they are fully ripe, though.) Harvest European pears when their background color changes slightly and they separate readily from the plant, then ripen them until soft and buttery in a cool (65° to 70° F) room. Pick medlars as the tree is shedding its leaves, then place the fruits on a shelf indoors to soften.

Once your fruit plants mature, you will undoubtedly begin to harvest more fruits than you can eat at one sitting, so you will have to either give away the surplus, or preserve it or store it fresh. You can preserve your surplus by drying, canning, and freezing it and by making jams and jellies.

Fresh fruits are a real treat anytime of the year, and with careful attention to types and varieties of fruits, as well as storage conditions, you could eat fresh fruit almost year round. Fresh fruits store best when kept at high humidity and at temperatures just above freezing. Cool temperatures slow the ripening of slightly underripe fruits, the aging of fully ripe fruits, and the growth of decay-causing microorganisms. High humidity and cold temperatures slow water loss from fruits, preventing shriveling. A perforated plastic bag or a plastic storage container with its lid slightly ajar maintains close to the optimum 90 percent humidity for fruits stored in the refrigerator. In fall and winter you may find other areas around your home—a garage, an unheated foyer or room, or a basement—that provide the needed cold storage temperatures if you run out of refrigerator space.

Remove fruit from cold storage sometime before you are ready to eat it. A pear that was picked underripe for storage may need to finish ripening at room temperature. But even fruit that is already ripe should be allowed to reach room temperature before you eat it so that you can appreciate its full flavor.

| Fruit | Days of storage possible under good conditions |
|---|---|
| Apple | 90–140 |
| Apricot | 7–14 |
| Blackberry | 2–3 |
| Blueberry | 14 |
| Cherry (sour) | 3–7 |
| Cherry (sweet) | 14–21 |
| Currant | 7–14 |
| Elderberry | 7–14 |
| Fig | 7–10 |
| Gooseberry | 14–28 |
| Grape (vinifera) | 90–180 |
| Grape (American) | 15–56 |
| Jujube | 365 |
| Juneberry | 2–3 |
| Kiwi | 270 |
| Medlar | 60 |
| Mulberry | 2–3 |
| Nectarine | 14–28 |
| Pawpaw | 90 |
| Peach | 14–28 |
| Pear | 60–210 |
| Persimmon | 90–120 |
| Plum | 14–28 |
| Pomegranate | 14–28 |
| Quince | 60–90 |
| Raspberry | 2–3 |
| Strawberry | 5–7 |

# Chapter 6

# GARDENERS' MOST-ASKED QUESTIONS

*Q:* **What fruit plants are especially ornamental, useful for food and beauty?**

*A:* *The showy blossoms of many fruit trees—especially peaches, apricots, plums, and cherries—are dazzling in spring. After flowering, the lustrous, drooping leaves of peaches* carry on the show until fall. Pomegranates bear flame-red flowers in spring and are semi-evergreen where winters are mild. Some fruit plants have been grown as ornamentals. Juneberries are valued for their pink or white flowers in spring as well as their good fall color. Kiwifruit vines make handsome arbor or pergola plants, with their apple green leaves, sometimes livened with white or pink variegations.

**Q: Can I use any fruit plants as hedges?**

A: Suckering shrubs that are good for hedges include pomegranates (dwarf pomegranates if you want a low hedge), black currants, Nanking cherry, and blueberries.

**Q: My fruit trees arrived this spring while the soil was still too wet to plant. How can I hold the plants until the soil is ready?**

A: Plants are usually shipped dormant and leafless in spring. Keep the roots moist until you are ready to plant by temporarily setting the roots in a shallow hole on the north side of a building or by temporarily potting the plant. Keep a temporarily potted plant in an unheated garage, where the cool temperature will delay bud break.

**Q: Why does my old apple tree bear two completely different types of apples, one of which tastes awful?**

A: Long ago your tree sent up a root sucker, which is a shoot that arose below the graft, from the rootstock. Rootstocks, when allowed to fruit, generally yield poor fruit. Trace the limb bearing poor fruit to its origin and cut it off.

**Q: Why are my apples all dimpled and riddled inside with brown trails?**

A: The symptoms described are common on neglected trees east of the Rockies and are caused by tunneling by apple maggot larvae, appropriately nicknamed "railroad worms." Use red spheres coated with sticky Tangletrap to attract females before they lay eggs, which hatch into "worms" in the fruit. Hang the spheres in trees about six weeks after bloom and leave in place for the whole season.

**Q: Do fruit plants have to be sprayed?**

A: Not necessarily. Whether or not you have to spray depends on what fruits you grow and where you live. Many common fruits can be grown in California without spraying. East of the Rockies, however, fruits such as apples, plums, and cherries need spraying. Bush fruits usually do not need any spraying no matter where you live. Grapes, gooseberries, and peaches are fruits that may or may not need to be sprayed, depending on the weather and site conditions.

**Q: Will any fruits bear in the shade?**

A: Most fruit plants need full sunlight—six or more hours per day—to stay healthy and productive, but some that bear well in partial shade include gooseberry, currants, hardy kiwifruit, persimmon, elderberry, and pawpaw.

**Q: Can't I just plant a peach or apple seed to start a peach or apple tree?**

**A:** Peach and apple seeds will grow into peach and apple trees, but the fruits on such trees will be of unpredictable quality, depending on what pollen fertilized what egg cells. Especially with apples, which need cross-pollination, the chances of a seedling tree bearing fruit even as good as that from the fruit that furnished the seed are very slim. Seedling trees are also slow to come into bearing—ten years or more might be needed for an apple tree to bear its first fruits when grown from seed.

**Q: Where can I go if I want more detailed information about potential pests as well as varieties specific to where I live?**

**A:** Your local Cooperative Extension service, with personnel in almost every county, can answer questions as well as provide publications for your region.

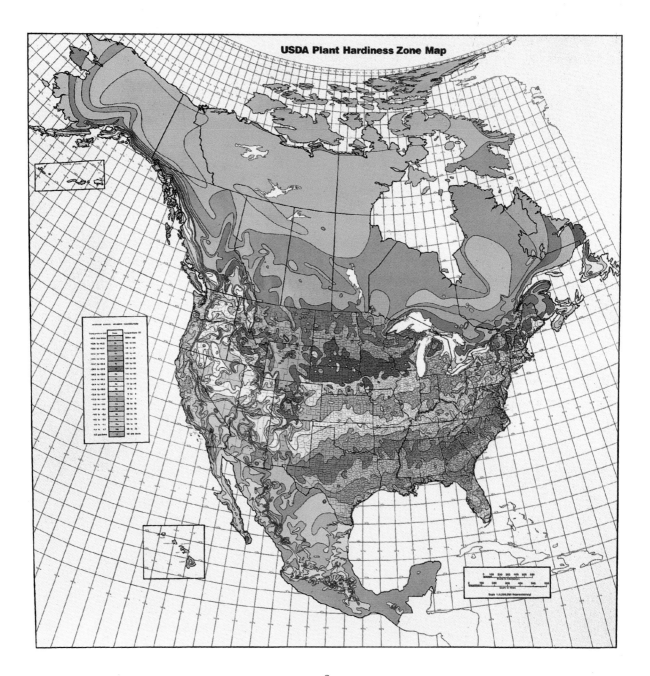

USDA Plant Hardiness Zone Map

# Index

# INDEX